The Imperial College Lectures in
# PETROLEUM ENGINEERING

An Introduction to
Petroleum Geoscience

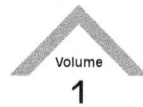

Volume
**1**

The Imperial College Lectures in
# PETROLEUM ENGINEERING

## An Introduction to Petroleum Geoscience

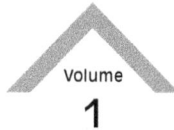

Volume
**1**

**Michael Ala**
Imperial College London, UK

**World Scientific**

NEW JERSEY · LONDON · SINGAPORE · BEIJING · SHANGHAI · HONG KONG · TAIPEI · CHENNAI · TOKYO

*Published by*

World Scientific Publishing Europe Ltd.

57 Shelton Street, Covent Garden, London WC2H 9HE

*Head office:* 5 Toh Tuck Link, Singapore 596224

*USA office:* 27 Warren Street, Suite 401-402, Hackensack, NJ 07601

**Library of Congress Cataloging-in-Publication Data**
Names: Ala, Michael, author.
Title: An introduction to petroleum geoscience / by Michael Ala, (Imperial College London, UK).
Description: [Hackensack] New Jersey : World Scientific, [2017] |
   Series: The Imperial College lectures in petroleum engineering ; volume 1 | Includes index.
Identifiers: LCCN 2016049017 | ISBN 9781786342065 (hc : alk. paper)
Subjects: LCSH: Petroleum--Geology. | Petroleum reserves.
Classification: LCC TN870.5 . A528 2017 | DDC 553.2/8--dc23
LC record available at https://lccn.loc.gov/2016049017

**British Library Cataloguing-in-Publication Data**
A catalogue record for this book is available from the British Library.

Desk Editors: Dipasri Sardar/Mary Simpson/Shi Ying Koe

Typeset by Stallion Press
Email: enquiries@stallionpress.com

# Preface

Petroleum geoscience is a multidisciplinary subject, underpinned by the broad spectrum of the Earth sciences. The aim of this book is to cover the fundamentals of the Earth sciences and examine their role in controlling the global occurrence and distribution of oil and gas resources. It also explains the principles, practices and the terminology associated with the upstream sector (exploration and production activities) of the oil industry. The book is intended also for non-geoscientists — petroleum engineers, drilling engineers and petrophysicists — who, due to their interaction with geologists and geophysicists, benefit from some knowledge of the latters' role and work practices in exploration ventures.

With the non-geoscientists in mind, the book opens with a chapter introducing the principles of geology, defined as the study of rocks and Earth history. The shape, structure, age and composition of the Earth are reviewed, followed by a brief discussion of the origin and description of the rocks comprising the crust, which contains the world's commercially accessible mineral and fossil fuel resources. Since hydrocarbons are hosted primarily in sedimentary rocks, emphasis is placed on their formation, evolution and descriptions of the most common types. Movements within the crust cause the rocks to deform. These deformational features are reviewed briefly in the closing sections of Chapter 1.

Worldwide, petroleum accumulations are found in *sedimentary basins* — depressions in the Earth's crust containing a large thickness of sedimentary rocks — and the parameters that determine hydrocarbon occurrence are geologically controlled. Plate tectonics plays a paramount role in explaining a wide variety of the Earth's geological and geophysical features, including the origin and subsequent evolution of sedimentary basins. An overview of the concept and its influence on basin formation, depositional history and characteristics is presented in Chapter 2. Petroleum is a chemically complex substance and encompasses both crude oil and gas. The chemical composition of petroleum and the impact on its physical properties of the relative proportions of its constituent components are discussed in Chapter 3.

Oil and gas fields are economically viable concentrations of hydrocarbons, the formation of which depends on a number of factors referred to as *elements* and *processes*. The study of how these elements and processes interact to create hydrocarbon-bearing provinces is known as *Petroleum system analysis*. Such analysis underpins the assessment of the hydrocarbon potential of the area under exploration and a major part of this book, Chapter 4, is devoted to the discussion of the generation, migration and accumulation of petroleum.

Exploring for oil and gas requires a systematic approach. Exploration techniques are described in Chapter 5. The first step in the assessment of the hydrocarbon potential of an area is the collection of data. Broadly, the information required falls into three categories: geological, remote sensing (obtained from aerial photography and satellite imagery) and geophysical (acquired by conducting gravity, magnetic and seismic surveys). Methods employed in the interpretation of these data are presented and discussed. Preparation of subsurface maps is an integral part of data interpretation. A large variety of such maps are in use and serve as indispensable aids in hydrocarbon reserves estimation. Exploration is inherently risky and the level of risk associated with drilling objectives identified during the exploration phase needs to be assessed and quantified. Evaluation of hydrocarbon potential and assessment of risk at basin or regional level is referred to as play fairway analysis.

Chapter 6 addresses the question of *resources* and *reserves*. The former is a broad term, encompassing all industrially useful materials, which include mineral deposits as well as petroleum, while the latter refers to the portion of an identified resource that can be produced economically through existing technology. Following a classification of the term reserves, methods of estimating the volumes of hydrocarbons initially in place in a reservoir are demonstrated and statistics are provided on the distribution of oil and gas reserves, production and consumption in global terms. Biofuels are also discussed.

It has been known for nearly two centuries that vast amounts of oil and gas are locked up in dark-coloured, organic-rich shale rocks. These are referred to as *unconventional hydrocarbons*, defined as oil and gas deposits that are not stored in pore spaces in permeable rocks (the reservoir) and are not commercially recoverable by "conventional" drilling and production methods. Advances in drilling and extraction technology since the mid-1990s, however, have made it possible for some of these hydrocarbons to be produced economically, resulting in their status being changed from unconventional to conventional. Prominent among these are shale oil, shale gas and oil or tar sands. In particular, oil and gas supplies from shale have risen rapidly in the US since the turn of the present century and have transformed the American energy landscape; the US is now self-sufficient in gas and its imports of crude oil have diminished significantly. The various types of unconventional hydrocarbons, their global distribution, estimates of the reserves attributed to them and production of oil and gas from these sources are reviewed in Chapter 7.

Oil and gas wells are usually cased. Casing generally protects the borehole wall, prevents it from collapse in unconsolidated formations, and shuts off the flow of unwanted fluids into the borehole. Hydrocarbons are produced through perforations in the production section of the casing. Before a well is cased, however, measurements are made of the electric, acoustic, radioactive, dielectric and nuclear magnetic properties of the rocks penetrated in the well. This is referred to as *well logging* and the interpretation of the measurements provides valuable information on the physical properties of sub-surface rock layers and their interstitial fluids as well allowing the calculation of fluid saturations. Knowledge of hydrocarbon saturation is indispensable in the computation of reserves. There are also logging tools that record the angle of inclination and dip direction of the layers traversed by the well and devices that provide computer-generated images of the borehole wall based on the electrical properties or the acoustic reflectivity of the formations. The acquisition and basic interpretation of wireline log data are dealt with in Chapter 8.

Finally, there is a glossary that defines the technical terms and explains the abbreviations used in the Earth and petroleum geosciences and charts the history and functions of some of the scientific and data gathering organisations that are associated with the international oil industry.

# About the Author

Dr. Michael Ala has been closely associated with the oil industry for more than 35 years as an exploration geologist with an independent oil company, as a consultant as well as in the field of education and training. He has a BSc in Oil Technology and an MSc, PhD and DIC in Petroleum Geology from Imperial College.

In 1973, he joined Seagull Exploration International and was involved in exploration studies and prospect evaluation in many parts of the world including Africa, northwest Europe, eastern Mediterranean, the Caribbean, South America and the Middle East. In 1976, he became the company's General Manager in London, responsible for its north European and Middle Eastern operations.

He joined the academic staff of the Earth Science Department of Imperial College in 1981, rising to the post of Director of the internationally recognised MSc Petroleum Geoscience Course in

1994. Dr. Ala has published more than 60 research and review articles covering the Middle East and West Africa, focusing on the petroleum geology and oil industry of Iran and is on the Editorial Board of the international *Journal of Petroleum Geology*. He was also Editor in Chief of *Seventy-Five Years of Progress in Oil Field Science and Technology*, published in 1990. Since 1982 he has been involved in organising and presenting numerous industrial training programs in Europe, throughout Africa, Middle East and southeast Asia.

Dr. Ala remains engaged in the oil industry and currently his commercial activities include providing advice on upstream projects and serving as a non-executive director on some oil company management boards.

# Acknowledgements

The material synthesised in this book has been drawn from many sources and its completion would not have been possible without the assistance of many people. A large number of the illustrations come from published literature and permission for their reuse had to be obtained from the copyright holders. It was necessary to redraw many of the figures and I am grateful to Caroline Baugh for organising the teams who carried out these tasks. Obtaining permissions from the copyright owners proved to be a complex and time consuming process and I am deeply indebted in this regard to Akash Kumar, Tarik Saif and Tom Dray. Their dedication, perseverance and patience in negotiating with the copyright proprietors were instrumental in securing the required consents.

Redrawing resulted in significant improvements in the quality of the illustrations concerned. This task was undertaken by Sarah Dodds, Christopher Dean and Bhavik Lodhia to whom I would like to express my gratitude and appreciation. Special thanks are due to my colleagues John Cosgrove for contributing the photographs shown in Figs. 1.14, 1.47, and 5.2 as well as producing the diagram in Fig. 7.7; Chris Jackson for supplying the seismic line presented in Fig. 5.28; and Al Fraser for providing the seismic section and its interpretation depicted in Fig. 5.30.

Last but not least, I would like to record my thanks to Imperial College Press for giving me the opportunity to write this book and to Tom Stottor and Mary Simpson, desk editors at World Scientific Publishing, for their helpful advice during its preparation.

# Contents

# Book Description

This book covers the fundamentals of the Earth sciences and examines their role in controlling the global occurrence and distribution of hydrocarbon resources. It explains the principles, practices and the terminology associated with the upstream sector (exploration and production activities) of the oil industry. Key topics include a look at the elements and processes involved in the generation and accumulation of hydrocarbons (petroleum system analysis) and demonstration of how geological and geophysical techniques can be applied to explore for oil and gas. There is detailed investigation into the nature and chemical composition of petroleum, coverage of surface and subsurface maps, including their construction and uses in upstream operations. Other topics include well-logging techniques and their use in determining rock and fluid properties, definitions and classification of resources and reserves, conventional oil and gas reserves, their quantification and their global distribution as well as unconventional hydrocarbons, their worldwide occurrence and the resources potentially associated with them. Finally, practical analysis is concentrated on the play concept, play maps, and the construction of petroleum events charts and quantification of risk in exploration ventures.

# Introduction to the Imperial College Lectures in Petroleum Engineering

This collection of volumes is based on the MSc in Petroleum Engineering course at Imperial College London. Each volume contains a single topic, or related collection of topics, based on the lectures given to our students. The purpose of the series is to provide an integrated overview of Petroleum Engineering as provided in our one-year intensive programme. The books provide somewhat more detail and discussion than written lecture notes, but are not intended to be comprehensive textbooks with a wide survey of the literature. The initial publication of four volumes will be complemented later by other contributions to develop a comprehensive, integrated whole.

Petroleum teaching and research at Imperial College began in 1913 with the appointment of Vincent Illing and the establishment of the BEng Oil Technology course. Over the last 102 years activities have evolved with the industry we serve, and we now offer a modern and rigorous education with an emphasis on the application of fundamental concepts to solve practical problems. We started MSc teaching in Petroleum Engineering in 1975 in response to the start of oil production from the North Sea. The course has grown over the intervening 40 years to reflect the international nature of the business. We offer a course which converts students with a good background in physical science or engineering into petroleum engineers ready to develop complex hydrocarbon resources using the latest ideas and technology.

The current structure of the MSc dates from 1997 with the appointment of Prof. Alain Gringarten as course director. One important theme was, and continues to be, the integration of engineering with geoscience, with common courses and project work conducted with students from our sister MSc course in Petroleum Geoscience. One important objective of both MSc courses is to educate students in the work flow concepts now used widely in the oil industry. We graduate petroleum professionals who are specialists in their respective fields, but who are trained to work effectively in multi-disciplinary teams. This follows from the recognition that the reservoir management process requires close cooperation between many different disciplines. We have implemented a curriculum that reflects the inter-disciplinary nature of how petroleum professionals work. Our MSc courses are hosted in the Royal School of Mines by the Department of Earth Science and Engineering: the department resides in the Faculty of Engineering combining earth science research and teaching with related engineering activities in mining, petroleum and computational modelling. The same seamless integration is seen within all aspects of our petroleum teaching and research.

Our MSc course teaches: (1) the fundamental concepts of reservoir characterisation, reservoir modelling, reservoir simulation, and field management; (2) the links between the various types of data; and (3) the processes for integrating and processing all available information. Our lectures, with related field work, are held in the first eight months of the course.

Formal lectures are complemented with a group field development project. This project concerns the evaluation of an oilfield using real data provided by operating oil companies. Initially this was from the Maureen Field in the North Sea, operated by Phillips Petroleum, and now is based on the Wytch Farm Field off the South coast of England which was operated by BP. The project has three phases which are distributed throughout the first six months of the programme. In the first phase, engineers and geoscientists work together in teams with six members to characterise a field, develop a reservoir model and estimate reserves with the associated uncertainty. In the second phase, a reservoir simulation model is developed and used in combination with analytical techniques to predict dynamic reservoir behaviour and propose a putative development plan. The final phase incorporates the design of production facilities, and economic and environmental considerations into an integrated proposal for development. The best two development projects are presented to an industry audience upon completion at the end of March. The best project presentation, as determined by the audience, is awarded the Colin Wall prize, in memory of Professor Colin Wall, who taught petroleum engineering at Imperial from 1964 to 1992.

Finally, the students complete an individual research project, taking approximately four months, usually based in industry, where new methods and ideas in various aspects of petroleum engineering are explored. The MSc thesis is written in the format of a paper for a Society of Petroleum Engineers (SPE) conference and indeed many of the students do later present and publish their work with the SPE.

## The Reservoir Management Process

Reservoir management is the application of available technology and knowledge to a reservoir system in order to control operations and maximise the economic value of the reservoir. It is about making the best possible decisions in line with the company's objectives. The ability to make the best possible reservoir management decisions relies mainly on the ability to predict the consequences of implementing these decisions and to evaluate the associated uncertainties. This in turn depends on the ability to model the expected behaviours of the reservoir system. The complete process, schematically represented in Fig. 1, includes four stages: reservoir characterisation, reservoir performance, well performance, and field development. This is mirrored in our teaching. Full details are shown in Fig. 2.

The first stage (shown in blue in Figs. 1 and 2), reservoir characterisation, corresponds to the identification of a model for the reservoir, the dynamic behaviour of which must be as similar as possible to that of the real field. It involves two consecutive steps: (1) the identification of data interpretation models; and (2)

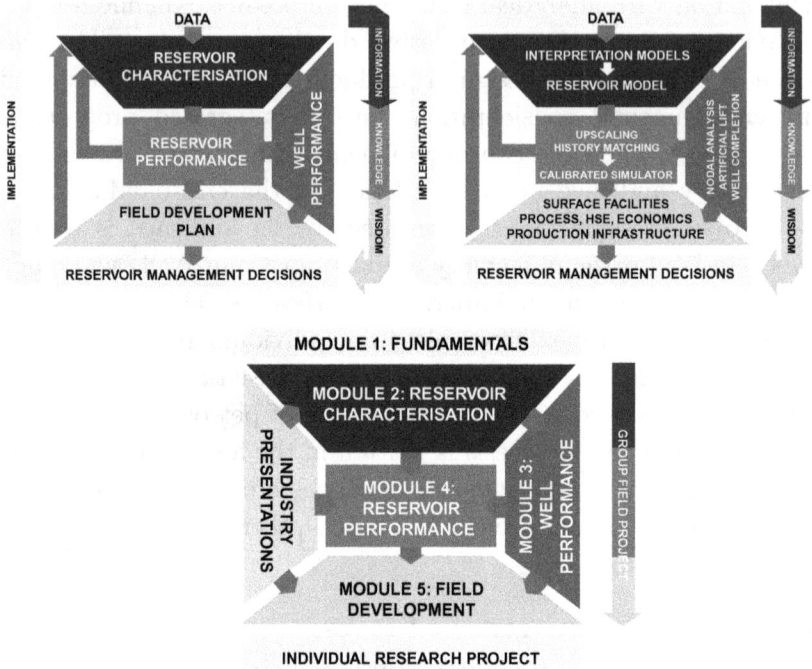

Figure 1. The reservoir management process in industry (top) and how this is mirrored in our MSc teaching (bottom). Figures taken from SPE 64311 "A Petroleum Engineering Educational Model Based on the Maureen Field UKCS", by Gringarten A. C., Bond D. J., Jackson M. D., Jing X., Ala M. and Johnson H. D.

the integration of these interpretation models into a reservoir model. Data interpretation models are obtained from the various types of reservoir data, i.e. "static" data such as geology, geophysics, geochemistry, and petrophysics which correspond to a description of the reservoir structure; and "dynamic" data, such as fluid pressures, geomechanics, tracers test results, production logs, well tests, and production, which are related to the flow behaviour of that reservoir. The integration of the data interpretation models into a reservoir model can be performed using either deterministic or stochastic approaches. Once the reservoir model is constructed, one must verify, as far as possible, that it is consistent with all available information and interpretation models. If the diagnosed reservoir

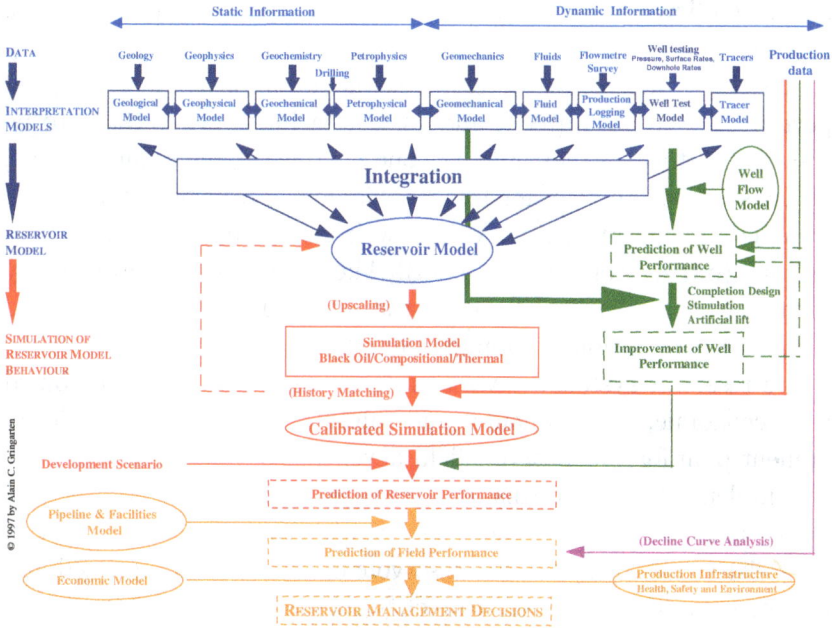

Figure 2. Details of the reservoir management process: the colours correspond to the four stages of our MSc teaching, reflected in Fig. 1. Figure taken from SPE 64311 "A Petroleum Engineering Educational Model Based on the Maureen Field UKCS", by Gringarten A. C., Bond D. J., Jackson M. D., Jing X., Ala M. and Johnson H. D.

model is consistent, the match between field data (seismic, logs, well tests, and production data) and the corresponding calculated model responses should be reasonably good and, in many cases, should be easy to improve by adjusting some reservoir model parameters within limits imposed by available knowledge (history matching). A lack of match which cannot be easily resolved may indicate an inconsistent or incomplete reservoir model and requires reinterpretation of the data. If the reservoir model is incorrect or incomplete, it is unlikely that optimum development will be achieved.

Once a consistent reservoir model is obtained, it can be used to predict future behaviours for various development scenarios (reservoir performance, shown in red). The production behaviour of the reservoir (pressure, rates, and saturations) is calculated with the calibrated flow simulator obtained as part of the verification process. In order to predict the behaviour of the field, however, it is necessary to consider the entire system, i.e. the reservoir, the wells (well performance in dark green) and surface facilities. These have to be treated together, as they influence one another: at any point, or node, of the system, the rate, for instance, is determined by equilibrating what is arriving at this node (inflow from upstream) with what is leaving (intake from downstream). The rate can be improved by acting either on the inflow or on the intake until an optimum and cost effective solution is obtained. Simulation of the behaviour of the entire reservoir system for various development scenarios combined with economic, environmental and safety constraints yield a development plan for the reservoir field development (shown in orange).

Implementation of this plan creates new data. The reservoir model must be tested for consistency with this new information. If inconsistencies are found, the reservoir model must be updated and the whole process repeated, to rectify the development plan.

The MSc has five modules taught over a six-month period which mirrors the steps described above: Fundamentals (Introductory Geology, Petroleum Geology, Petroleum Geophysics, Rock Properties, Fluid Properties, Flow in Porous Media, Production Mechanisms, Drilling); Reservoir Characterisation (Geological Modelling, Geophysical Modelling, Geochemistry, Geomechanics, Petrophysics, Production logging, Well Testing, and Integration into a Reservoir Model); Well Performance (Completions and Production Problems, Nodal Analysis); Reservoir Performance (Analytical Predictors, Numerical Predictors, Upscaling, History Matching, Enhanced Oil Recovery and Practical use of numerical simulators), and Field Development (Pipelines and Surface Facilities, Environment and Safety, Economics).

Teaching is provided by approximately equal numbers of Imperial College faculty members and industry specialists for those disciplines which are not present at Imperial. Courses are organised sequentially, i.e. students receive teaching in only one discipline at any given time. Presentations by industry experts and careers talks are provided every week. The field development group project runs parallel

Figure 3. Integration from the start. All 134 of our MSc students in Petroleum Engineering and Petroleum Geoscience are shown on a field trip to Dorset during the first two weeks of the course.

to the lectures, with deliverables corresponding to the respective modules (i.e. reservoir model, calibrated simulation model, and field development plan).

We make a concerted effort to promote close cooperation between the MScs in Petroleum Geoscience and Petroleum Engineering: Students and faculty members from these disciplines are located in a common area for project work, and the Fundamental and Reservoir Characterisation modules are taught largely in common. This includes a five-day geological field trip, see Fig. 3. Most "static" courses are now taken jointly, while geoscientists are given a shorter version of the Petroleum Engineering "dynamic" courses. The objective is to create a common culture, so that professionals from the various disciplines can understand their own potential contribution and that of others to the reservoir characterisation process. To that end, the reservoir characterisation phase of the group project is organised with multidisciplinary groups from the two courses, thus reproducing the environment of an asset team.

Our Petroleum Engineering MSc is very successful, attracting over 500 applicants for only 50–60 students each year. All our students have the equivalent of a first class honours degree while approximately half have relevant industrial experience. We take students from a variety of backgrounds from all over the world. Typically, over 20 different countries are represented in each student cohort while in 2015, 37% of our students were women. Enrolment is deliberately limited to allow every student to receive an excellent education with time to discuss problems with individual lecturers.

While it cannot reproduce the Socratic nature of detailed questioning combined with engineering practice through project work and research as part of the MSc course, we trust that this series of volumes will help teach Petroleum Engineering to students everywhere, particularly those who are not fortunate enough to be at Imperial.

Prof. Martin Blunt took over from Prof. Gringarten as course director in 2013. We anticipate that the course will continue to flourish, while developing to meet new challenges.

We would like to thank our colleagues who have worked so hard in the preparation of these volumes, and our many students, now employed around the world, whose curiosity and insights have helped us all become better teachers. We trust that these books will help educate, inspire and stimulate a new generation of engineers in the industry.

Martin Blunt and Alain Gringarten
London, April 2015

# Chapter 1

# Introduction to Geology

## 1.1. Introduction

Geology is the study of rocks and Earth history. The Earth is a dynamic and ever changing body; earthquakes, volcanoes, moving continents and expanding and contracting oceans are all surface phenomena providing evidence of an active interior. They also provide clues to the internal processes, structure and composition of the Earth.

A unique feature of the Earth is that it has liquid water. Figure 1.1 presents a view of the Earth from space.

## 1.2. Shape, Internal Structure, and Composition of the Earth

In terms of its shape, the Earth is an oblate sphere; it is relatively flat at the poles but bulges at the equator due to the centrifugal force acting on it as the result of its rotation. Consequently, its polar radius (6,357 km) is 21 km shorter than its equatorial radius (6,378 km). A mean value of 6,371 km is often given for the radius of the Earth.

Figure 1.2 shows a section through the Earth, illustrating its internal structure and composition.

The inner core is solid but the outer core is thought to be liquid. The evidence for this is, however, indirect and is based on the study of the propagation of seismic waves generated by earthquakes. Seismic energy travels through the Earth in the form of compressional

Figure 1.1. The Earth as seen from space (http://s3.amazonaws.com/estock/
nasas1/1/81/98/everystockphoto-nasa-space-18198-o.jpeg).

and shear waves, normally referred to as P-waves and S-waves,
respectively. The modes of propagation of these waves are different.
Considering the Earth as consisting of particles, P-waves propagate
by causing the particles to vibrate (oscillate) horizontally and the
direction of motion is also horizontal, as illustrated in Fig. 1.3.
S-waves, by contrast, travel by causing the particles to vibrate
(oscillate) vertically, while the direction of propagation is horizontal,
as shown in Fig. 1.4. The velocity of both P- and S-waves increases
with depth, as can be seen in Fig. 1.5. It should be noted that P-wave
velocity, $V_p$, is greater than that of the S-waves, $V_s$:

$$V_p = 1.7 V_s. \tag{1.1}$$

Earthquakes generate both P- and S-waves, which travel through
the Earth and can be detected at observation stations all around

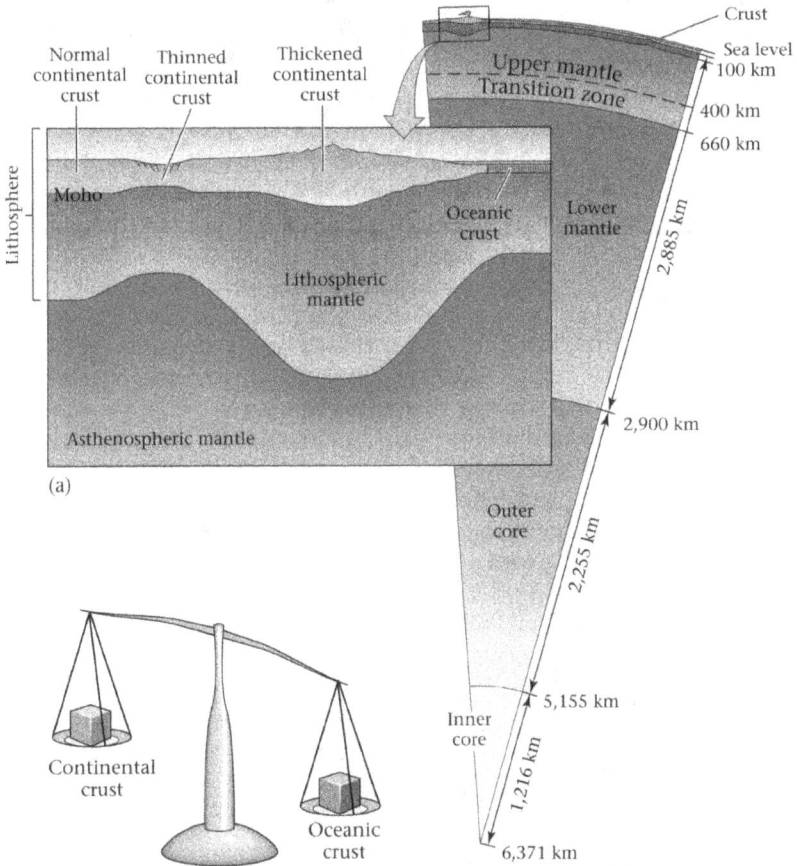

Figure 1.2. Structure and internal composition of the Earth (after Marshak, 2005).

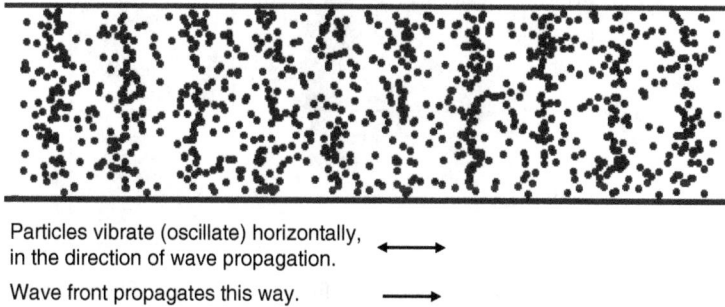

Particles vibrate (oscillate) horizontally, in the direction of wave propagation. ←→

Wave front propagates this way. →

Figure 1.3. P-wave transmission.

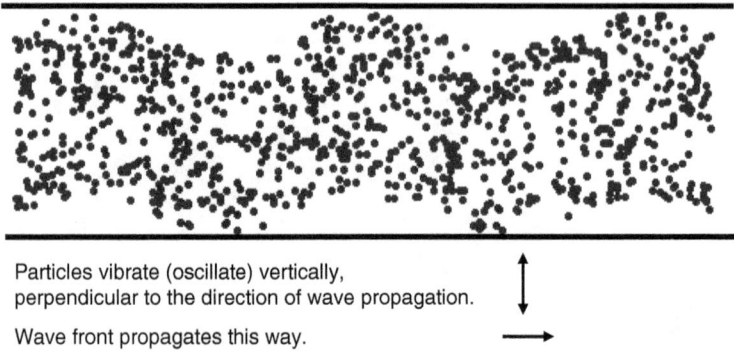

Particles vibrate (oscillate) vertically,
perpendicular to the direction of wave propagation.

Wave front propagates this way.

Figure 1.4.  S-wave transmission.

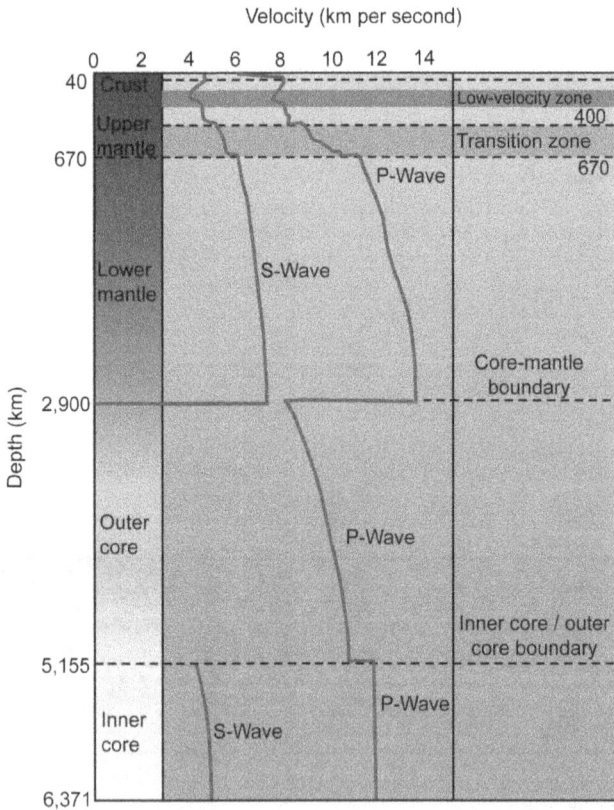

Figure 1.5.  Seismic wave velocity/depth profile for the Earth. There is a general
increase in velocity with depth. No S-waves pass through the outer core, indicating
that it is liquid (after Marshak, 2005).

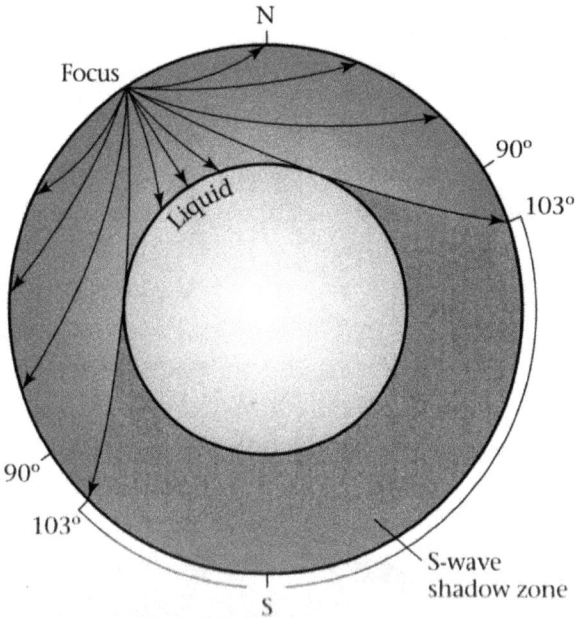

Figure 1.6. The S-wave "shadow zone". This covers more than a third of the globe (after Marshak, 2005).

the world. They can be separated due to the difference in their velocities; on account of their higher velocity, the P-waves constitute the first arrivals. In order to reach stations in the S-wave "shadow zone" indicated in Fig. 1.6, the waves would have to pass through the outer core. Only P-wave arrivals are detected at these stations, which means that the S-waves have been absorbed in the course of their passage through the outer core. Transmission of S-waves is a property of solids; liquids are unable to transmit S-waves, which supports the conclusion that the outer core is liquid.

The crust, containing all our mineral deposits and fossil fuel resources, forms only 0.6% of the Earth's radius. It is divided into oceanic and continental types, with the latter being denser (Fig. 1.2). Crust thicknesses vary from 40 km in continental areas to 10 km in oceanic regions. Oxygen, silicon and aluminium dominate the composition of the crust (Fig. 1.7), while the whole Earth composition is dominated by iron and oxygen (Fig. 1.8). The circulating flow of

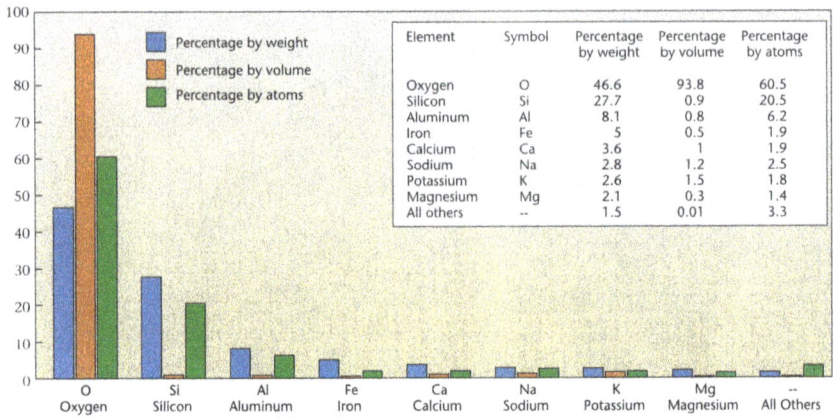

| Element | Symbol | Percentage by weight | Percentage by volume | Percentage by atoms |
|---|---|---|---|---|
| Oxygen | O | 46.6 | 93.8 | 60.5 |
| Silicon | Si | 27.7 | 0.9 | 20.5 |
| Aluminum | Al | 8.1 | 0.8 | 6.2 |
| Iron | Fe | 5 | 0.5 | 1.9 |
| Calcium | Ca | 3.6 | 1 | 1.9 |
| Sodium | Na | 2.8 | 1.2 | 2.5 |
| Potassium | K | 2.6 | 1.5 | 1.8 |
| Magnesium | Mg | 2.1 | 0.3 | 1.4 |
| All others | -- | 1.5 | 0.01 | 3.3 |

Figure 1.7. Abundance of elements in the Earth's crust (after Marshak, 2005).

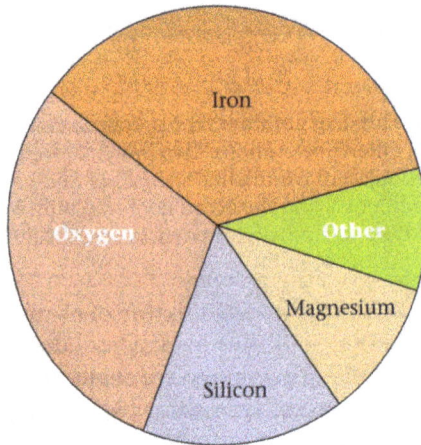

Figure 1.8. Whole Earth composition (after Marshak, 2005).

the liquid in the outer core is the cause of the Earth's magnetic field.

## 1.3.  How Old is the Earth?

Until the 18th century, the Bible was the source of knowledge on virtually all subjects. Archbishop Ussher, head of the Anglo-Irish

Church in Ireland, by adding up the generations of the patriarchs described in the Old and New Testaments, concluded in 1654 that the "Earth came into being on Sunday 23 October, 4004 BC" (Marshak, 2005).

The first scientific attempt at estimating the age of the Earth was by Kelvin, the Scottish physicist, in the 1890s. He calculated the time for the Earth to cool down from an original molten state, as hot as the sun, to its present temperature and concluded that this was about 20 million years.

The discovery of radioactivity by the French physicist Becquerel in 1896 revolutionised thinking on the age of the Earth. It led to the realisation that the Earth possessed an internal source that had generated heat; radioactive decay was identified as the source of this heat in the course of geological time. This realisation uncovered the flaw in Kelvin's calculation: he had assumed that no heat was added to the planet after it was formed.

Current estimates place the age of the Earth at 4.66 billion years and the oldest rocks encountered to date are in Canada. These rocks have been dated at 4 billion years by radiometric age determination methods (see Sec. 1.11.1).

## 1.4. The Earth's Crust (Lithosphere)

As mentioned earlier, the crust or lithosphere contains all of the Earth's mineral deposits and fossil fuel resources. The crust is made up of *rocks*, which may be defined as aggregates of crystals (or of non-crystalline materials) or grains. Geologists recognise three basic rock types:

- igneous,
- metamorphic,
- sedimentary.

These are described briefly in Table 1.1.

Igneous and metamorphic rocks lack hydrocarbon prospects and are referred to as "economic basement" in petroleum geological terminology. They are of high-temperature origin, while sedimentary

Table 1.1. A brief description of crustal rocks.

| | |
|---|---|
| **Igneous rocks** — form by the solidification of *magma* or molten material that rises from the interior of the Earth. These are the 'parent' rocks of the other 2 groups. Example: *granite*.<br><br>**Metamorphic rocks** — form by the alteration of pre-existing rocks through an increase in temperature and pressure. The alteration occurs in the solid state, which means that it does not involve melting. Examples: *marble and slate*. | 'Economic Basement' non-petroliferous |
| **Sedimentary rocks** — form either by the cementing together or consolidation of fragments (grains) derived from pre-existing rocks, by direct precipitation from water, or from the life processes of animals and plants. Examples: *sandstone, limestone, shale, chalk*. | Petroliferous |

rocks form under low-temperature (surface) conditions. Oil and gas are associated with **sedimentary** rocks.

## 1.5.  Formation of Igneous Rocks

Igneous rocks solidify from *magma*, molten material that forms in chambers below the Earth's surface, as illustrated in Fig. 1.9. Broadly, they fall into the *extrusive* and *intrusive* types. In the case of the former, the magma reaches the surface through vents, forming volcanic cones and moving down the mountainside as lava flows (Fig. 1.10). The lava then cools and forms solid igneous rocks. Not all of the magma finds its way to the surface; some of it forces its way into the surrounding rocks of the crust and, as it cools, forms horizontal and vertical or sub-vertical bodies, referred to as *sills* and *dykes* respectively. These bodies constitute the intrusive igneous rocks.

Igneous rocks are the most abundant rock types on Earth; they make up all of the mantle, all of the oceanic crust and much of the continental crust. Their colour ranges from pale grey and pink to black, depending on the chemical composition of the parent magma.

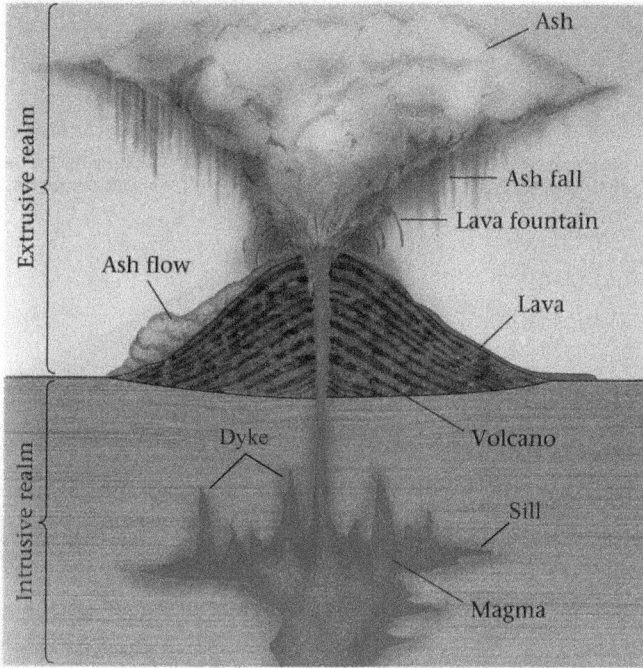

Magma collects in a subsurface "chamber" and gives rise to
extrusive and intrusive igneous activity (after Marshak, 2001).

A sill exposed due to uplift and erosion, east Greenland
(from Schofield, 2015).

Figure 1.9. Formation of igneous rocks and a field photograph of a sill.

Figure 1.10. Volcanic eruption. Magma flowing from a vent at the surface — extrusive igneous activity.

*Source*: https://www.carbonbrief.org/wp-content/uploads/2015/

Figure 1.11 shows hand specimens of some of the most common igneous rocks, namely granite and basalt.

## 1.6.  Formation of Metamorphic Rocks

The action of heat and pressure on pre-existing rocks, regardless of their type, leads to the formation of metamorphic rocks. Temperature and pressure increase as the result of burial and proximity to or intrusion by igneous bodies and cause changes in the mineralogical composition of the rock; new minerals are formed at the expense of the old ones. There are also changes in the shape, size and arrangement of the grains or crystals in the rock. The degree of alteration (*metamorphic rank*) is determined by the intensity of the heat and pressure to which the original rock has been subjected. Metamorphic changes take place in the solid state and the process does not involve melting. Figure 1.12 presents examples of metamorphic rocks.

(http://www.beg.utexas.edu/mainweb/publications/graphics/granite.htm)

Granite hand specimen
Acidic igneous rock (rich in silica, $SiO_2$)

Basalt hand specimens
Basic igneous rock (rich in iron and
magnesium)

Figure 1.11. Examples of igneous rocks. Rock type and colour are determined by the chemical composition of the original magma. Silica rich magmas produce light-coloured igneous rocks while magmas rich in iron and magnesium give rise to dark-coloured igneous rocks.

(http://www.geol.ucsb.edu/faculty/hacker/geo102C/lectures/bonsVein.jpg)
Metamorphic rock outcrop (gneiss)

Gneiss hand specimen

Figure 1.12. Examples of metamorphic rocks. Note the extensively fractured and intruded nature of the outcrop: evidence of the rocks subjected to high pressure and temperature.

## 1.7. The Rock Cycle

The transformation of Earth materials from one rock type to another is called the *rock cycle*. It illustrates the relationship among the three rock types briefly described in Table 1.1. The transformation is progressive and involves a series of interrelated events that have occurred frequently in the geological history of the Earth.

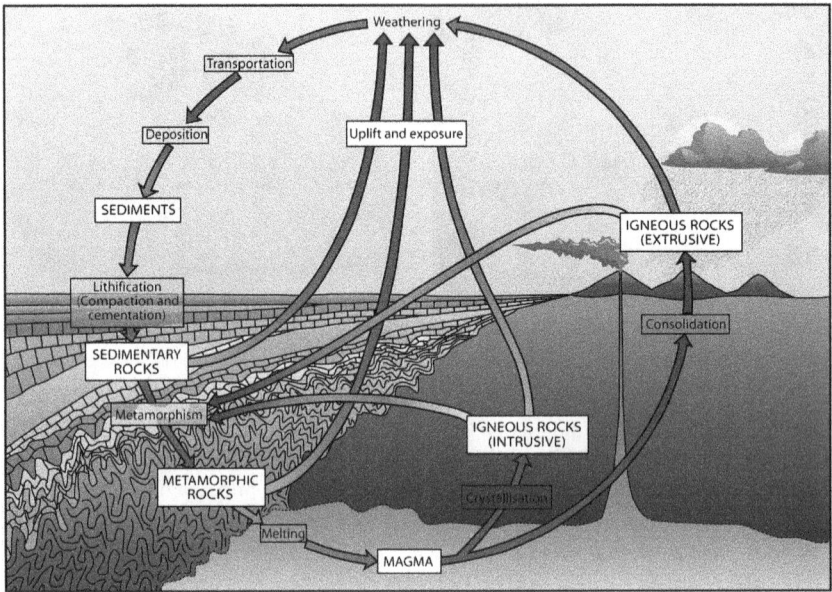

Figure 1.13. The rock cycle, illustrating the relationship among the rock types (modified and redrawn from The Rock Cycle).

Figure 1.13 is an illustration of the rock cycle. The most complete path is around the circle and can be symbolised as igneous $\Rightarrow$ sedimentary $\Rightarrow$ metamorphic $\Rightarrow$ igneous. Magma forms in the subsurface and reaches the Earth's surface via volcanic eruptions. At the surface, it solidifies into igneous rocks, which become subjected to weathering and erosion, producing particles that are transported by running water, wind and ice from the source area to the site of deposition where they undergo lithification and form sedimentary rocks. Upon burial, the sedimentary rocks are subjected to heat and pressure and form metamorphic rocks. Melting occurs as the result of further burial and increase in temperature, leading to the formation of magma. However, the path does not always follow this sequence of events; the metamorphic rocks could be uplifted and eroded to form new sediments and new sedimentary rocks without melting and turning into igneous rocks, thus taking a "shortcut" through the cycle; the igneous rocks may be metamorphosed directly without going through the sedimentary part of the cycle; sedimentary rocks may be subjected to uplift and erosion directly, forming new sediments and new sedimentary rocks without undergoing metamorphism, melting, magma generation and the igneous rock formation phases of the cycle.

## 1.8.   Sedimentary Rocks: Key Features and Terminology

*Bedding* or *stratification* is the most important distinguishing feature of sedimentary rocks. This is clearly evident in the examples shown in Fig. 1.14 and demonstrates that they were deposited as discrete layers. The layers or *strata* are separated by *bedding planes*.

### 1.8.1.   *Formation*

Broadly, sedimentary rocks are formed through the following processes:

- Consolidation of individual particles or grains derived from the erosion/breakdown of pre-existing rocks and minerals. Figure 1.15 illustrates the production of grains from a granite fragment. Deposits formed in this manner are called *clastic* or *detrital* rocks. Typical examples: *sandstone* and *shale*.
- Direct precipitation from seawater. These are referred to as *chemical* rocks. Typical examples: *limestone and evaporite*.
- Life activities of animals and plants. These are known as *biochemical* rocks. Typical example: *coral reefs*.

The formation of clastic rocks involves transportation of the grains or fragments from their source to a place of deposition through the agencies of water, wind and ice. Chemical and biochemical rocks are found largely where they are formed, i.e. they are *in situ* deposits.

### 1.8.2.   *Sedimentary environment*

The *sedimentary environment* is defined as the physical and chemical conditions that prevail in an area where sedimentary rocks are being deposited. As illustrated in Fig. 1.16, there are many different types of environment; they can exist at the same time and each one

(Photo by M. Ala)

(Photo by J. Cosgrove)

Figure 1.14. Exposures of sedimentary rocks showing their bedded or stratified nature.

is characterised by a particular deposit or *facies* (Fig. 1.17) that characterise a rock unit.

Facies represent the record of a sedimentary environment that existed in the geologic past but is no longer present. The rock type can, therefore, be used to reconstruct the environment at the time of its deposition. Sedimentary environments are classified into

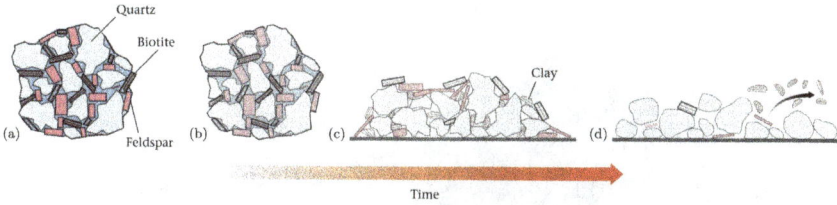

Figure 1.15. Formation of grains by breakdown of a pre-existing rock fragment, in this case granite. The less stable minerals, feldspar and biotite in this case, decompose and are removed. The grains of quartz (silicon dioxide, $SiO_2$), a stable mineral, remain and consolidate to form a sandstone (after Marshak, 2005).

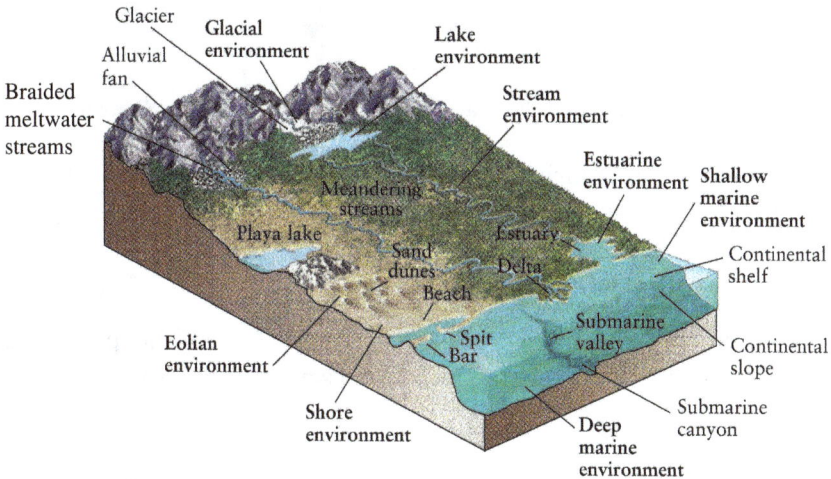

Figure 1.16. Illustration of sedimentary environments (after Murck & Skinner, 1999).

continental, shoreline and marine types, as illustrated in Fig. 1.18, which also mentions the processes and climatic conditions that prevailed at the time of their existence.

### 1.8.3. *Fauna and flora*

*Fauna* and *flora* refer to the animal and plant fossil remains, respectively, found in some sedimentary rocks. They are important indicators of the original environment of deposition, e.g. sub-aqueous

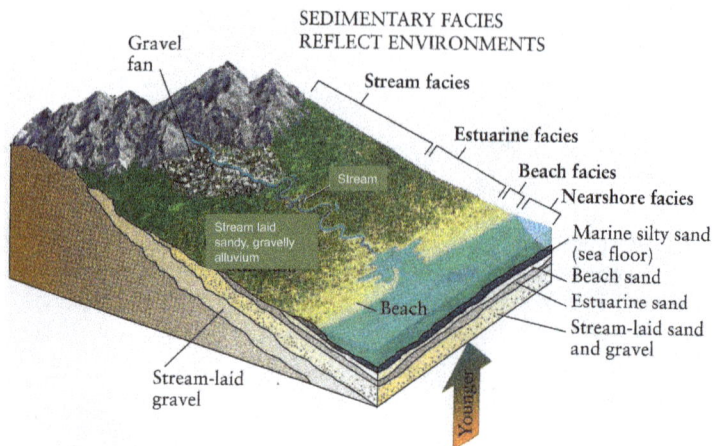

SEDIMENTARY FACIES
REFLECT ENVIRONMENTS

Gravel fan

Stream facies

Estuarine facies

Beach facies

Nearshore facies

Stream

Stream laid sandy, gravelly alluvium

Marine silty sand (sea floor)

Beach sand

Estuarine sand

Stream-laid sand and gravel

Beach

Stream-laid gravel

Younger

Figure 1.17. Definition and illustration of facies (after Murck & Skinner, 1999).

versus sub-aerial; marine versus non-marine; deep versus shallow water; humid versus arid climatic conditions.

## 1.9.   Common Sedimentary Rocks

The common sedmentary rocks and their place in the petroleum system are listed in Table 1.2 and are described briefly below.

### 1.9.1.   *Sandstone*

*Sandstone* is a clastic rock, formed by the consolidation of individual sand size particles (1/16–2 mm in diameter). Grains are transported from their source to the site of deposition, which means that sandstone is not an *in situ* deposit. As depicted in Fig. 1.19, clastic grain size decreases with increasing distance from the source area. Figure 1.20 presents the widely used Wentworth clastic rock grain size classification, and a sandstone hand specimen is shown in Fig. 1.21.

### 1.9.2.   *Carbonates*

*Carbonates* include limestone, made up of the mineral calcite $(CaCO_3)$, chalk and dolomite, which consists of the mineral dolomite $[MgCa(CO_3)_2]$. Limestones are either of chemical or biochemical origin and are formed largely *in situ*. Chemical limestones result

| ENVIRONMENT | SUB-ENVIRONMENT | TRANSPORT AGENT | SEDIMENTS | CLIMATE |
|---|---|---|---|---|
| Continental | Lacustrine (lake) ① | Lake currents and waves | Sand and mud, saline deposits in arid climates | Arid to humid |
| | Fluvial ② | River currents | Sand, mud and gravel | Arid to humid |
| | Desert ③ | Wind | Sand and dust | Arid |
| | Glacial ④ | Ice and meltwater | Sand, mud and gravel | Cold |
| Mixed or shoreline | Delta ⑤ | River currents waves | Sand and mud | Arid to humid |
| | Beach ⑥ | Waves and tidal currents | Sand and gravel | Arid to humid |
| | Tidal flats ⑦ | Tidal currents | Sand and mud | Arid to humid |
| Shallow Marine | Continental shelf ⑧ | Waves and tides | Sand, mud and carbonates carbonates and calcified organisms | |
| | Organic reefs ⑨ | Waves and tides | | |
| Deep marine | Continental margin ⑩ | Ocean currents and waves | Mud and sand | |
| | Deep sea ⑪ | Ocean and turbidity currents | Mud and sand | |

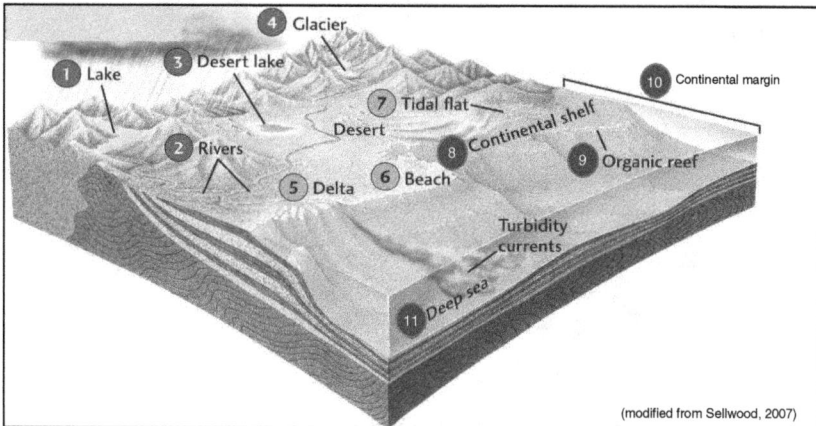

(modified from Sellwood, 2007)

Figure 1.18. Classification of sedimentary environments.

Table 1.2. Common sedimentary rocks.

| Name | Function in the petroleum system |
|---|---|
| Sandstones | Reservoir rocks |
| Carbonates | |
| Shales | Source and cap rocks (seals) |
| Evaporites | Cap rocks (seals) |
| Coal* | Potential source of gas (e.g. Southern North Sea) |

*Organic carbon, not a sedimentary rock, *sensu stricto*.

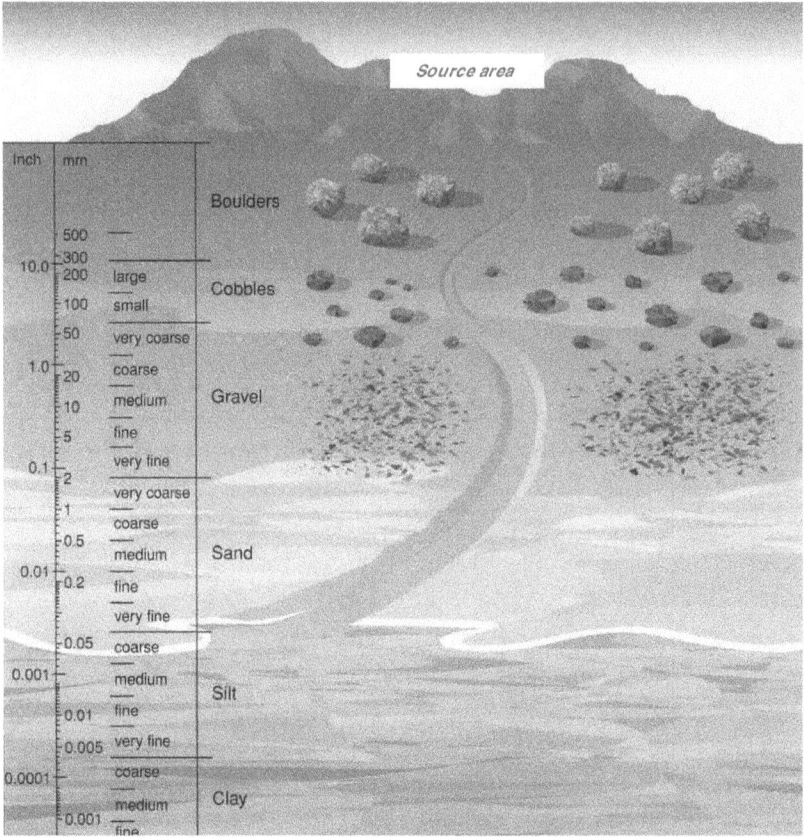

Figure 1.19. Decrease in clastic grain size with increasing distance of transport from source area (image copyright Schlumberger, used with permission and courtesy of Schlumberger).

from the direct precipitation of calcite from seawater, while the biochemical ones are the product of the life processes of living organisms such as corals, which build reefs, and algae. Figure 1.22 shows a hand specimen of fossiliferous limestone and Fig. 1.23 illustrates that reefs typically grow in shallow, warm, clear water environments. A reef rock specimen is presented in Fig. 1.24.

Chalk is a biochemical carbonate made up entirely of the remains of primitive, submicroscopic algae known as *coccoliths*. It is not as hard as limestone; it crumbles easily and hence is not a suitable building stone. It is characterised by a white to pale grey colour

| | |
|---|---|
| | Boulders |
| 256 mm | |
| | Cobbles |
| 64 mm | |
| | Pebbles |
| 4 mm | |
| | Granules |
| 2 mm | |
| | Very Coarse Sand |
| 1 mm | |
| | Coarse Sand |
| 1/2 mm | |
| | Medium Sand |
| 1/4 mm | |
| | Fine Sand |
| 1/8 mm | |
| | Very Fine Sand |
| 1/16 mm | |
| | Silt |
| 1/256 mm | |
| | Clay |

Figure 1.20. The Wentworth clastic rock grain size scales.

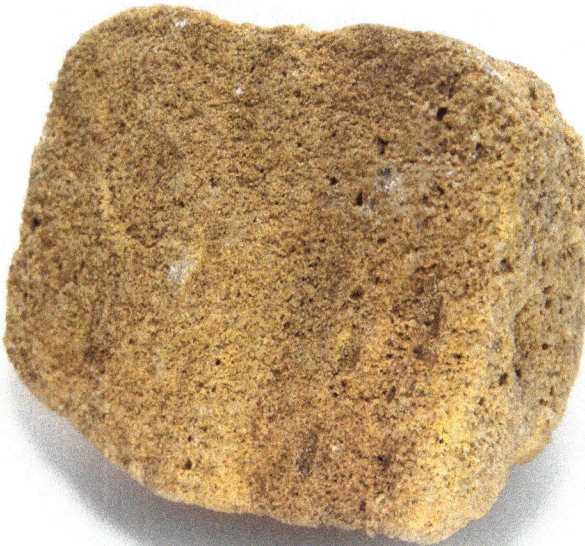

Figure 1.21. A sandstone hand specimen. Note the granular texture of the rock.

Figure 1.22. A fossiliferous limestone hand specimen (Virtual Geology Museum, Cochise College).

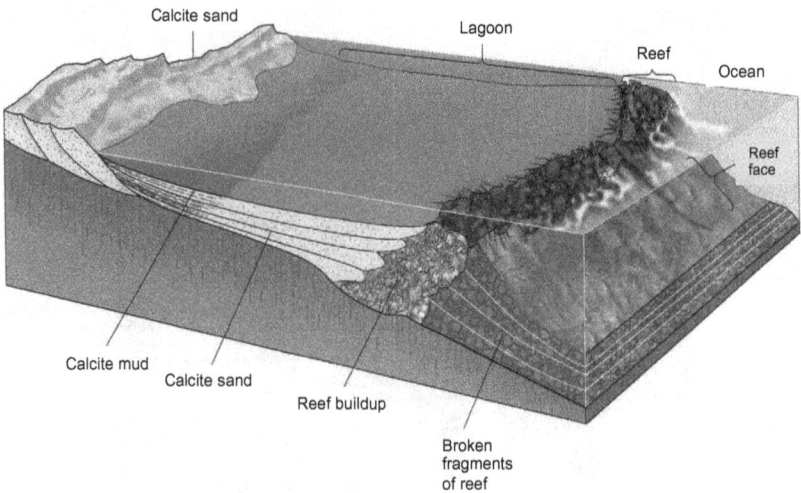

Figure 1.23. Reef environment (after Marshak, 2005).

and forms impressive cliffs along the southern coast of England (Fig. 1.25).

Whether dolomite originates through direct precipitation from seawater — similarly to calcite — is open to debate. Often it results

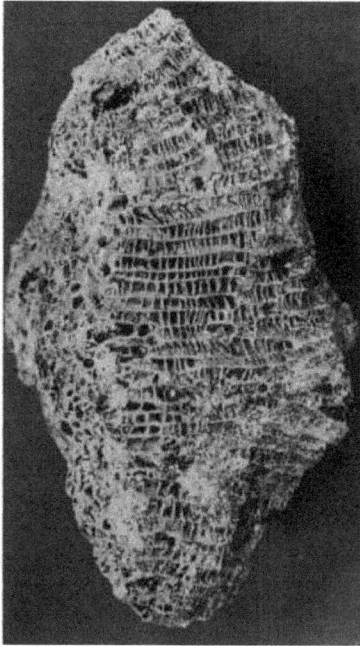

Figure 1.24. Reef rock specimen (from Harbaugh, 1965).

Figure 1.25. Chalk cliffs, southern England (http://i1.mirror.co.uk).

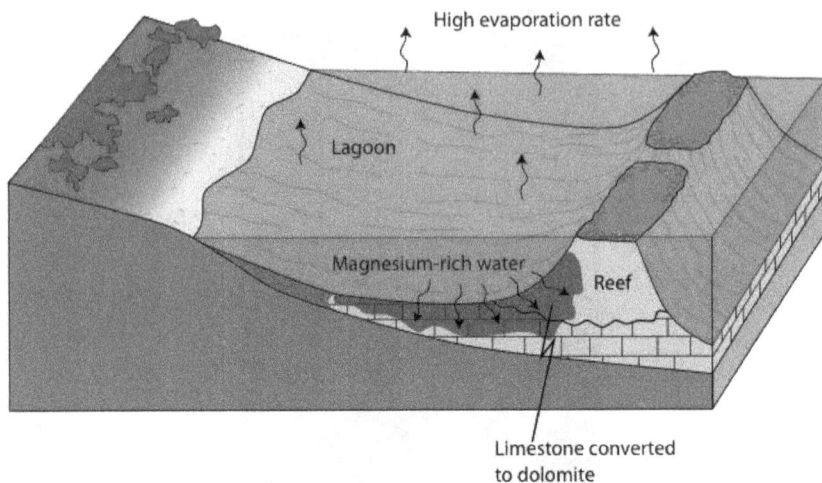

Under certain conditions, magnesium-rich waters percolate through the limestone, converting calcite into dolomite:
$MgCl_2 + 2CaCO_3 = Ca\,Mg\,(CO_3)_2 + CaCl_2$.

Figure 1.26. Mechanism of formation of dolomite (dolomitisation) (modified and redrawn from Sellwood, 2007).

from the conversion of calcite due to percolation by magnesium-rich waters as illustrated in Fig. 1.26. The process is referred to as *dolomitisation*.

### 1.9.3.  *Shale*

*Shale* is a very fine-grained clastic rock composed of silt and clay size particles (less than 1/16 mm in diameter). Due to their very low permeability, shales can act as cap rock or seals to oil and gas accumulations. Organic-rich shales, deposited in anoxic conditions, can act as petroleum source rocks. A photograph of a shale exposure is presented in Fig. 1.27.

### 1.9.4.  *Evaporites*

*Evaporites* are chemically precipitated from concentrated solutions or *brines*. They form in hot regions, where evaporation exceeds the inflow of water into the environment, and include anhydrite ($CaSO_4$), gypsum ($CaSO_4 \cdot 2H_2O$), halite ($NaCl$), sylvite ($KCl$) and bischofite ($MgCl_2 \cdot 2H_2O$), to name just a few. Evaporites are the most

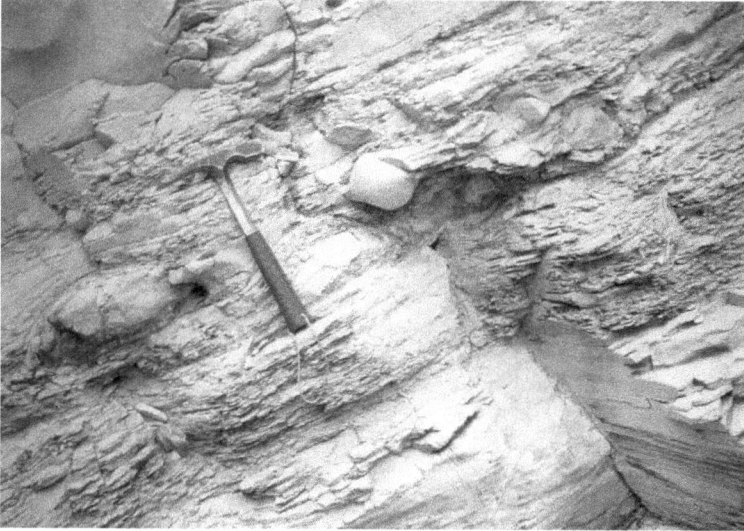

Figure 1.27. A shale outcrop. Note the laminated (fissile) nature of the rock (photo by M. Ala).

effective cap rocks. Figures 1.28 and 1.29 show specimens of halite and anhydrite respectively. They can be distinguished by taste.

### 1.9.5. *Coal*

*Coal* is formed by the burial and compaction of terrestrial plant remains. Deep burial causes degasification, producing gas. The gas in the southern North Sea fields was derived from the underlying coal measures by this process.

## 1.10. Lithification of Sediments into Sedimentary Rocks

The processes that are involved in the conversion of sediments into lithified sedimentary rocks are known collectively as *diagenesis*. Diagenesis begins immediately after deposition and results from chemical reactions between the rock and groundwater percolating through it, and rises in temperature and pressure as the depth of burial increases. It continues until the onset of metamorphism, which

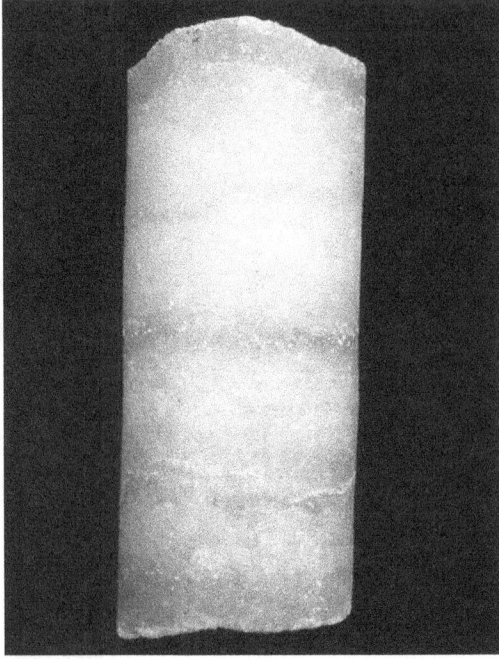

Figure 1.28. Rock salt (halite) (Photo by M. Ala).

Figure 1.29. A gypsum hand specimen.

is brought about by temperatures in excess of 200°C and elevated pressures as the result of deep burial.

Diagenetic processes are many, but the most important include compaction, cementation and replacement of existing crystals with new ones. These processes are described briefly below.

### 1.10.1.  *Compaction*

Sediments deposited sub-aqueously initially contain large volumes of water. This water is expelled as the grains undergo compaction due to the increase in pressure exerted by the growing weight of the overburden accumulating above. The grains become more closely packed and the density of the rock increases. Figure 1.30 presents a diagrammatic illustration of compaction.

### 1.10.2.  *Cementation*

Cementation involves the precipitation of crystalline material in the pore spaces between the grains by mineral-rich waters percolating

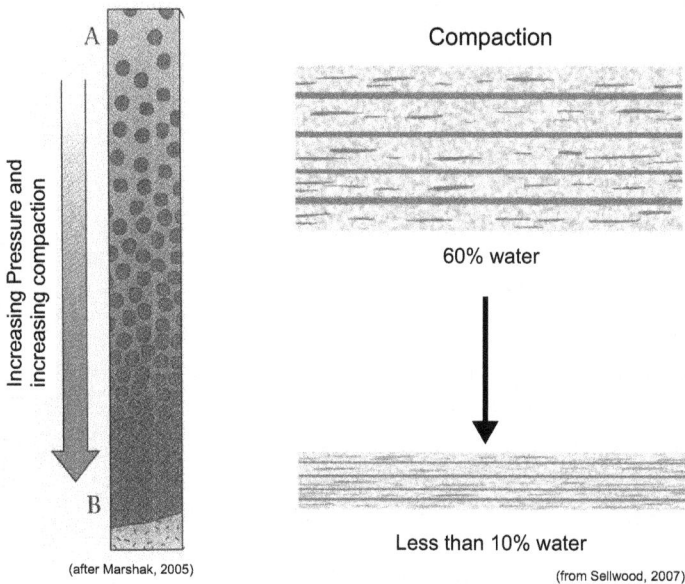

Figure 1.30. Illustration of compaction.

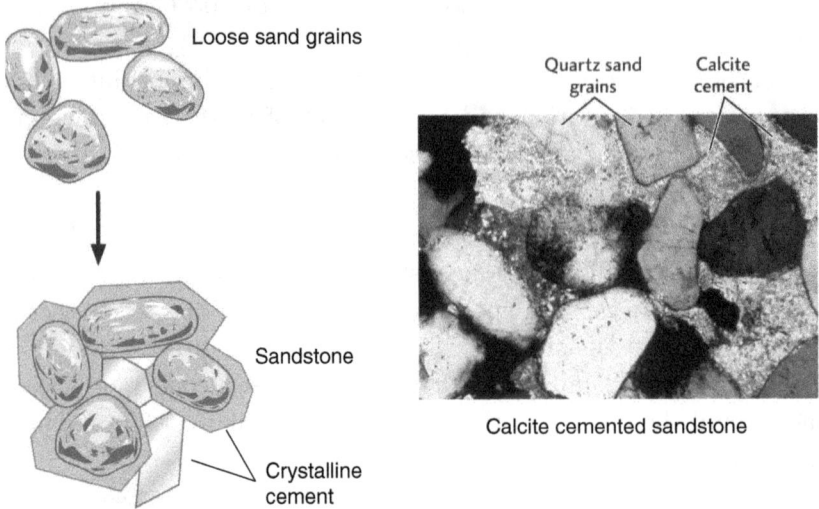

Figure 1.31. Illustration of cementation. Crystalline material precipitates in the pore spaces, causing the grains to consolidate into hard sandstone (from Sellwood, 2007).

through the rock. This is shown in Fig. 1.31. Calcite ($CaCO_3$) and silica ($SiO_2$) are the most common cements in sandstones.

Cementation reduces the original pore volume. However, since calcite is susceptible to solution, if after cementation the rock is subjected to percolation by ground waters charged with carbon dioxide, then the resulting carbonic acid dissolves the calcite cement and brings about an increase in pore volume.

### 1.10.3.  *Dolomitisation*

*Dolomitisation* is an example of the replacement of calcite crystals by dolomite crystals, resulting in the conversion of limestone into dolomite. It is considered as a diagenetic change caused by the reaction between calcite and magnesium chloride, which occurs in some percolating waters. The conversion takes place according to the following chemical reaction (see also Fig. 1.26):

$$MgCl_2 + 2CaCO_3 \Rightarrow CaMg(CO_3)_2 + CaCl_2.$$

## 1.11. Geological Time

Geological time may be expressed in relative or absolute terms. Until the discovery of radioactivity towards the end of the 19$^{th}$ century, the concept of absolute time was unknown and only relative ages of rock units could be determined. This was the practice followed by the pioneers who established the foundations of the science of geology in the late 18$^{th}$ century and throughout the 19$^{th}$ century. The concept of relative time is illustrated in Fig. 1.32.

### 1.11.1. *Absolute time*

Following the discovery of radioactivity in 1896, it was recognised early in the 20$^{th}$ century that certain radioactive elements decay

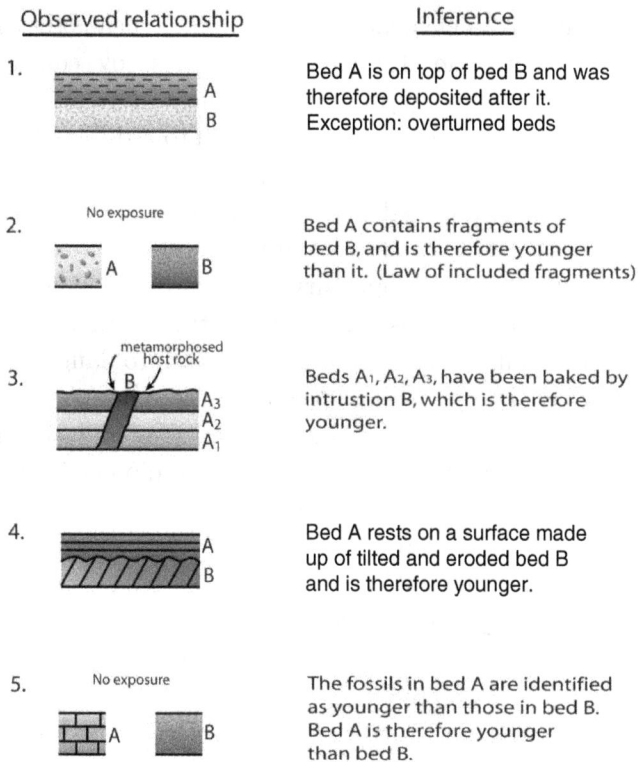

| Observed relationship | Inference |
|---|---|
| 1. | Bed A is on top of bed B and was therefore deposited after it. Exception: overturned beds |
| 2. No exposure | Bed A contains fragments of bed B, and is therefore younger than it. (Law of included fragments) |
| 3. metamorphosed host rock | Beds A$_1$, A$_2$, A$_3$, have been baked by intrusion B, which is therefore younger. |
| 4. | Bed A rests on a surface made up of tilted and eroded bed B and is therefore younger. |
| 5. No exposure | The fossils in bed A are identified as younger than those in bed B. Bed A is therefore younger than bed B. |

Figure 1.32. Relationships used to deduce relative ages of rock units.

with time and that the rate of this decay provided a means of measuring geological time in terms of years, i.e. absolute time. All atoms of a given element have the same number of protons and this is called its *atomic number*. The number of neutrons in the atomic nuclei varies, however, which means that not all atoms of a given element have the same *atomic weight*, defined as the total number of the protons and neutrons. This gives rise to different versions of an element, which are known as *isotopes* of that element. Isotopes are therefore characterised by the same atomic number but a different atomic weight. For example, the element uranium, with 92 protons that define its atomic number, has two isotopes, uranium-235 and uranium-238 — abbreviated to $^{235}$U and $^{238}$U, respectively — reflecting the variation in their atomic weight caused by differences in the number of neutrons in their nuclei.

Some isotopes of an element are stable, which means they remain unchanged through time. Radioactive isotopes, by contrast, are unstable, which means they undergo a change after a given time, as a result of which they become converted into a different element. This conversion is referred to as *radioactive decay* and the isotope that undergoes decay is called the *parent atom* while the product of decay is known as the *daughter atom*. It is not possible to specify how long a given radioactive isotope survives before it begins to decay, but what can be measured is the length of time it takes for half of the number of a group of parent atoms to decay into daughter atoms; this time is referred to as the *half-life* of the isotope.

The concept of a half-life is illustrated in Fig. 1.33 and half-life periods of the radioactive elements used in radiometric dating of rocks, and the minerals in which they occur, are listed in Fig. 1.34. Decay curves for $^{238}$U–$^{207}$Pb, $^{87}$Rb–$^{87}$Sr and $^{14}$C–$^{14}$N are presented in Figs. 1.35–1.37. The $^{14}$C–$^{14}$N method is used extensively in archaeological investigations. Parent-to-daughter isotope ratios in a mineral are determined in the laboratory and from this the age of the mineral can be calculated.

The half-life periods of various radioactive elements were worked out in the 1920s and 1930s and for the first time it became possible to assign absolute ages to the subdivisions of the geological column.

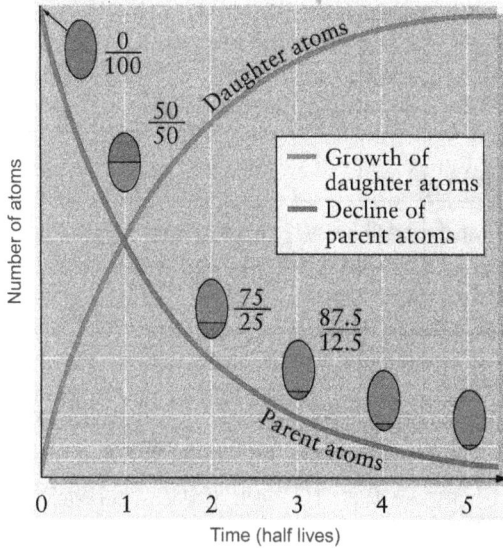

Figure 1.33. Radioactivity and time. The curves illustrate the decay of the *parent* element into the *daughter* element. The *half-life* is defined as the time taken for 50% of the parent atoms to decay into daughter atoms (after Murck & Skinner, 1999).

| Parent → Daughter | Half-Life (years) | Minerals in Which the Isotopes Occur |
|---|---|---|
| $^{147}Sm \rightarrow {}^{143}Nd$ | 106 billion | Garnets, micas |
| $^{87}Rb \rightarrow {}^{87}Sr$ | 48.8 billion | Potassium-bearing minerals (mica, feldspar, hornblende) |
| $^{238}U \rightarrow {}^{206}Pb$ | 4.5 billion | Uranium-bearing minerals (zircon, uraninite) |
| $^{40}K \rightarrow {}^{40}Ar$ | 1.3 billion | Potassium-bearing minerals (mica, feldspar, hornblende) |
| $^{235}U \rightarrow {}^{207}Pb$ | 713 million | Uranium-bearing minerals (zircon, uraninite) |

Sm = samarium, Nd = neodymium, Rb = rubidium, Sr = strontium, U = uranium, Pb = lead, K = potassium, Ar = argon

Figure 1.34. Isotopes and half-life periods of radioactive minerals used in the radiometric dating of rocks (after Marshak, 2005).

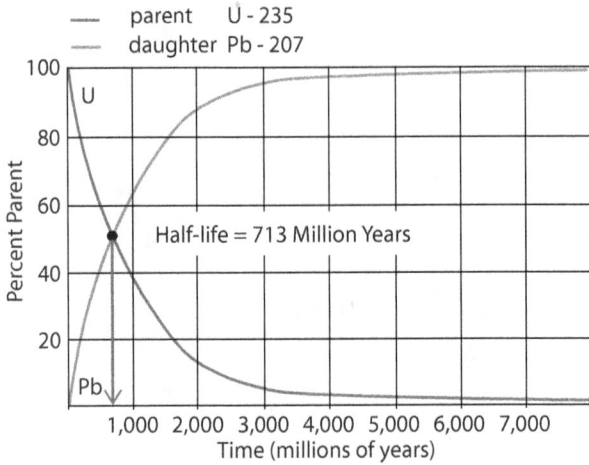

Figure 1.35.  $U^{235}$–$Pb^{207}$ decay curves.

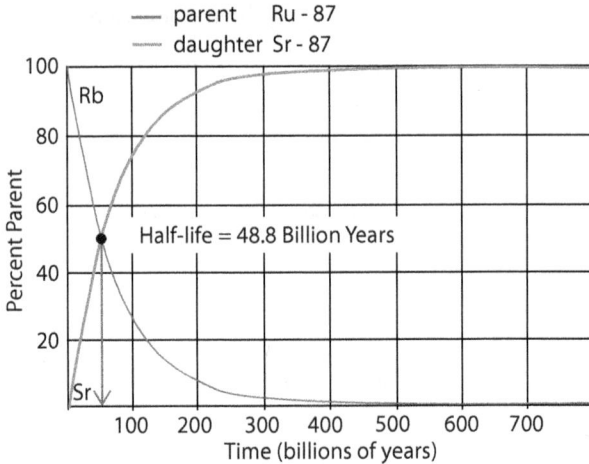

Figure 1.36.  $Rb^{87}$–$Sr^{87}$ decay curves.

Uncertainties in radiometric age determinations are of the order of 1% or less. For example, a date may be reported as $270 \pm 2.7$ million years. The age of the oldest rocks determined by radiometric dating is 4.2 billion years.

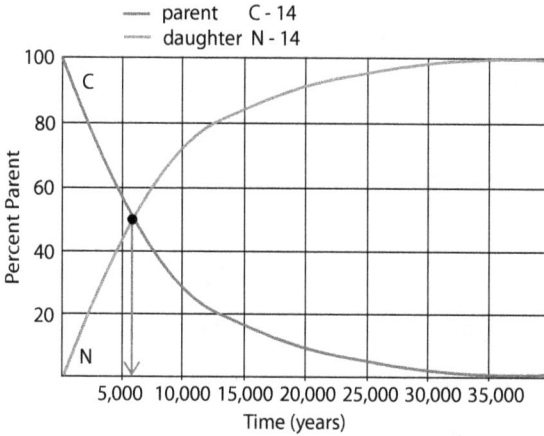

Figure 1.37. $C^{14}$–$N^{14}$ decay curves.

## 1.12. Stratigraphic Practice

*Stratigraphy* is a very important branch of geology, which is concerned with the study of the age and description of stratified rocks. Due to the immense span of geological time, it is necessary to divide it into manageable smaller portions. These smaller subdivisions form the *stratigraphic* or *geological column*. Figure 1.38 presents the major subdivisions of geological time and these are shown in more detail in Fig. 1.39. A summary of the evolution of life during the Phanerozoic Eon, the last 545 million years of Earth history, is presented in Fig. 1.40.

Stratigraphic practice is underpinned by several fundamental concepts, the most important of which can be summarised as follows.

- The *uniformitarian principle*. The present is the key to the past; i.e. the processes that operate at present also operated in the past and produced the same results.
- The *principle of superposition*. In a normal sedimentary sequence, i.e. one that is not upside down or inverted, the lowermost bed is always the oldest and topmost layer is always the youngest. This is demonstrated by pouring different coloured sand into a cylindrical jar, as shown in Fig. 1.41.
- The *principle of original horizontality*. The surfaces on which sediments accumulate — such as the floodplain of a river or the bed of a lake or sea — are approximately horizontal. A steep slope

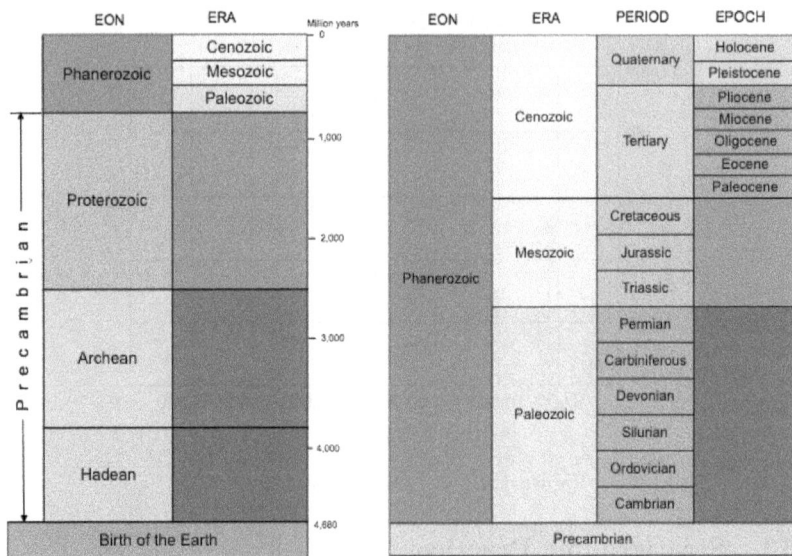

| EON | ERA | Million years |
|---|---|---|
| Phanerozoic | Cenozoic | 0 |
| | Mesozoic | |
| | Paleozoic | |
| Proterozoic | | 1,000 |
| | | 2,000 |
| Archean | | 3,000 |
| Hadean | | 4,000 |
| Birth of the Earth | | 4,680 |

Precambrian (vertical label on left)

| EON | ERA | PERIOD | EPOCH |
|---|---|---|---|
| Cenozoic | | Quaternary | Holocene |
| | | | Pleistocene |
| | | Tertiary | Pliocene |
| | | | Miocene |
| | | | Oligocene |
| | | | Eocene |
| | | | Paleocene |
| Phanerozoic | Mesozoic | Cretaceous | |
| | | Jurassic | |
| | | Triassic | |
| | Paleozoic | Permian | |
| | | Carbiniferous | |
| | | Devonian | |
| | | Silunan | |
| | | Ordovician | |
| | | Cambrian | |
| Precambrian | | | |

Figure 1.38. Major subdivisions of geological time.

would cause the sediments to slide downslope before lithification, which means they would not be preserved as sedimentary layers. Folds and tilted beds therefore indicate that the layers have been subjected to post-depositional deformation.

Sedimentary successions can be described in terms their fossil content (*biostratigraphic characteristics*) or their rock type or lithology (*lithostratigraphic characteristics*). These characteristics can be used to establish the relative ages of the rocks and trace their extension from one locality to another. Lithostratigraphy is more frequently used in petroleum geology, particularly in reservoir studies: reservoir characterisation and the establishment of flow units, i.e. identification of permeable layers that allow the transmission of fluids. In lithostratigraphic practice, the sedimentary successions are divided into *groups, formations* and *members.*

- A *formation* is defined is a rock unit which is different from the layers above and below by its lithological characteristics — mineralogy, colour, grain size, texture and structures — and has regional lateral continuity. Several formations that are somehow related make-up a *group.*

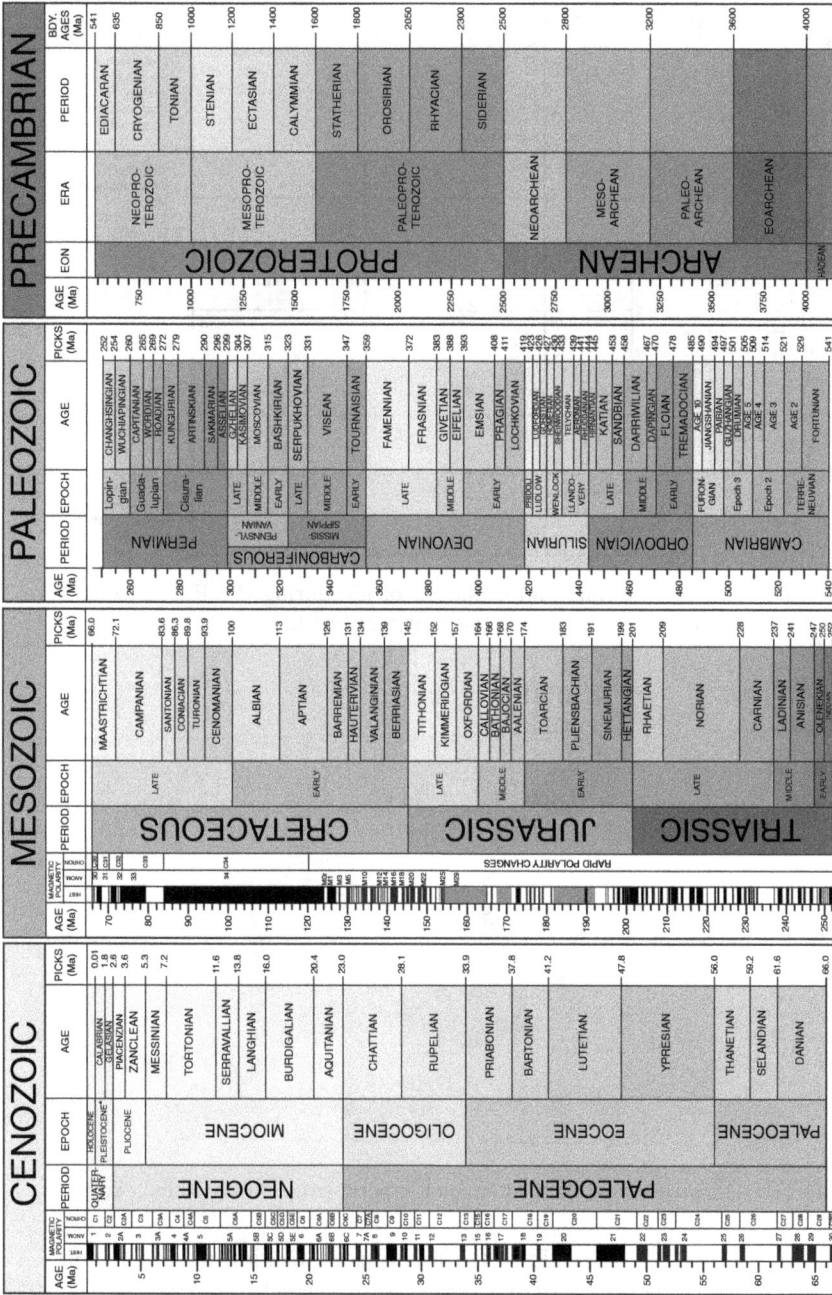

Figure 1.39. Subdivisions of geological time in greater detail (http://www.von-der-bank-online.de/stratigraphic_chart.jpg).

| PERIOD | | PLANT EVOLUTION | ANIMAL EVOLUTION |
|---|---|---|---|
| | Quaternary | Repeated glaciation | First *Homo sapiens* |
| | Tertiary | Decline of forests, spread of grasslands | First *Homo* (human) |
| | | | Apes and hominids appear |
| | | | All modern mammals represented |
| | | Explosive radiation of flowering plants | In seas, bony fish abound |
| | | | Rise of mammals |
| | Cretaceous | First flowering plants | Dinosaurs extinct |
| | | | Modern birds |
| | Jurassic | Forests of gymnosperms and ferns over most of the earth | First birds |
| | | | Age of dinosaurs |
| | Triassic | Gymnosperms dominant | Explosive radiation of dinosaurs |
| | | | First dinosaurs and mammals |
| | | | First beetles |
| | Permian | Decline of nonseed plants | Mammal-like reptiles appear |
| | | | Reptiles and insects increase |
| | | | Amphibians decline |
| | Pennsylvanian / Mississippian | Gymnosperms appear Widespread forests of giant club moss trees, horsetails and tree fern - create vast coal deposits | Early reptiles |
| | | | First winged insects |
| | | | Amphibians increase |
| | Devonian | First seed plants Development of vascular plants - club mosses and ferns | Amphibians diversify |
| | | | First land vertebrates -amphibians |
| | Silurian | First vascular plants First land plants | Golden age of fishes |
| | | | First land invertebrates -scorpions |
| | Ordovician | | First vertebrates - fishes |
| | | | Marine invertebrates increase |
| | Cambrian | Algae dominant | Trilobites dominate |
| | | | Explosive evolution of marine life |

Major extinction

Phanerozoic 541

2.59
66.0
145.0
201.3
252.2
298.9
358.9
419.2
443.8
485.4
541.0

Cenozoic / Mesozoic / Paleozoic
Proterozoic
2500
Archean
3800
Hadean
4600

☐ During the Proterozoic and the first 100m years of the Phanerozoic life was aquatic.

☐ The earliest terrestrial organisms were plants.

☐ Mammals, particularly humans, appeared very late in Earth's history.

Sudden appearance of abundant life forms. *This event is called the "Cambrian explosion".*

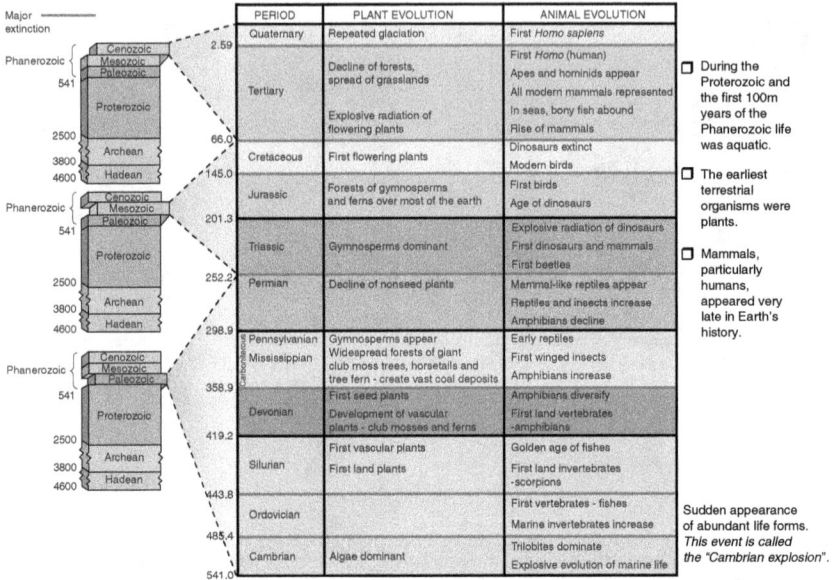

Figure 1.40. Geological time and evolution of life during the Phanerozoic Eon (modified and redrawn from Murck and Skinner, 1999).

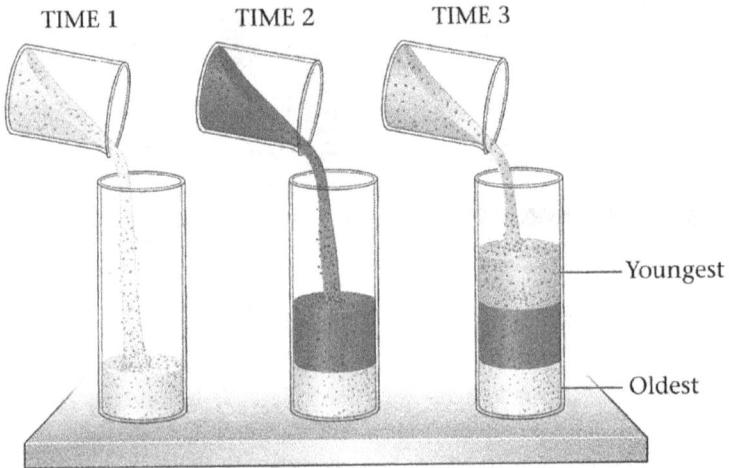

Figure 1.41. Illustration of the principle of superposition (after Marshak, 2005).

- A *member* is a subdivision of a formation. Regional lateral continuity is not required and it represents local lithological variations in the formation.

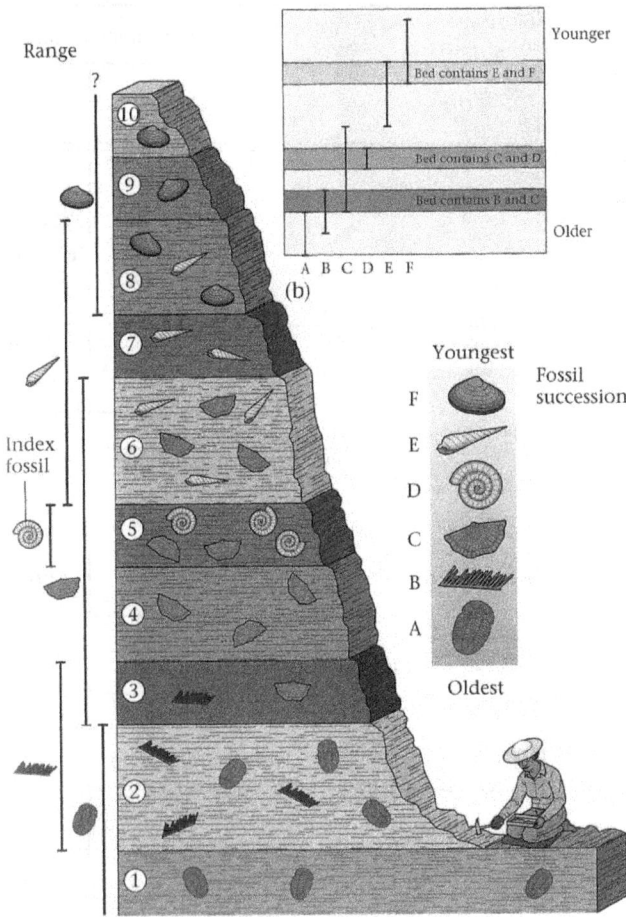

Figure 1.42. Principle of establishing biostratigraphic units. Note that each species (fossil type) has a limited vertical range and ranges of different fossils may overlap. Widespread species with a short range are index or zone fossils (after Marshak, 2005).

- A *bed* is lowest unit in terms of rank or hierarchy; it refers to an individual rock layer.

Figure 1.42 illustrates the principle of establishing biostratigraphic units by describing a sedimentary succession based on its fossil content, while the subdivision of a sequence into lithostratigraphic units is demonstrated in Fig. 1.43.

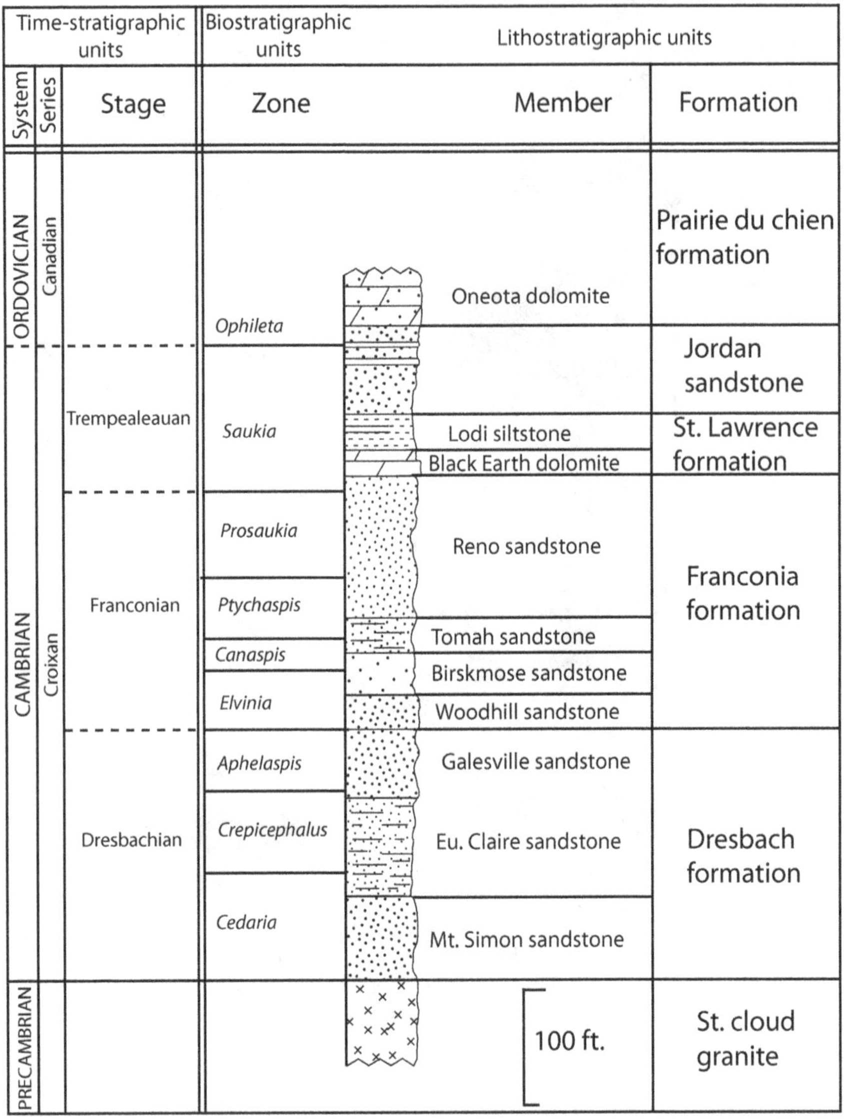

| Time-stratigraphic units | | | Biostratigraphic units | Lithostratigraphic units | |
|---|---|---|---|---|---|
| System | Series | Stage | Zone | Member | Formation |
| ORDOVICIAN | Canadian | | *Ophileta* | Oneota dolomite | Prairie du chien formation |
| CAMBRIAN | Croixan | Trempealeauan | *Saukia* | | Jordan sandstone |
| | | | | Lodi siltstone | St. Lawrence formation |
| | | | | Black Earth dolomite | |
| | | Franconian | *Prosaukia* | Reno sandstone | Franconia formation |
| | | | *Ptychaspis* | Tomah sandstone | |
| | | | *Canaspis* | Birskmose sandstone | |
| | | | *Elvinia* | Woodhill sandstone | |
| | | Dresbachian | *Aphelaspis* | Galesville sandstone | Dresbach formation |
| | | | *Crepicephalus* | Eu. Claire sandstone | |
| | | | *Cedaria* | Mt. Simon sandstone | |
| PRECAMBRIAN | | | | 100 ft. | St. cloud granite |

Figure 1.43. Subdivision of a sedimentary sequence into lithostratigraphic (formations and members) and biostratigraphic units. Note that biostratigraphic and lithostratigraphic unit boundaries do not always coincide. Fossil names are recorded in italics (modified and redrawn from Krumbein and Sloss, 1963).

## 1.12.1. *Stratigraphic relationships*

Lithological characteristics of sedimentary layers vary horizontally as well as vertically. Horizontal variations are called *lateral facies changes* (e.g. lateral gradation of a sandstone unit into a shale unit). Vertical relationships between successive units are said to be *conformable* if there has been no interruption in deposition between the two and *unconformable* if the layers are separated by a period of non-deposition or erosion. The surface of separation in the latter case is called an *unconformity*. Where there is an angular difference between the attitude of the overlying and underlying rocks, the surface of separation is called an *angular unconformity*.

Unconformities therefore represent time gaps in the sedimentary record — a period of variable length during which there was no deposition, or a removal of the rock record due to erosion. Figure 1.44

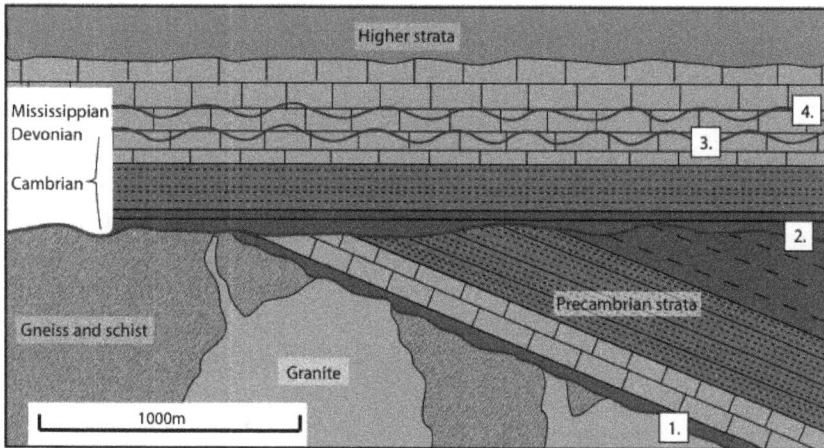

Figure 1.44. Illustration of unconformable relationships. Angular unconformities (1) and (2) represent large-scale crustal movements followed by long intervals of uplift and of erosion. These indicate large gaps in the sedimentary record. Surfaces (3) and (4) are almost parallel to the rock layers and represent broad uplift followed by some erosion. They indicate smaller gaps in the sedimentary record and are referred to by some authors as *disconformities* (modified and redrawn from Flint and Skinner, 1974).

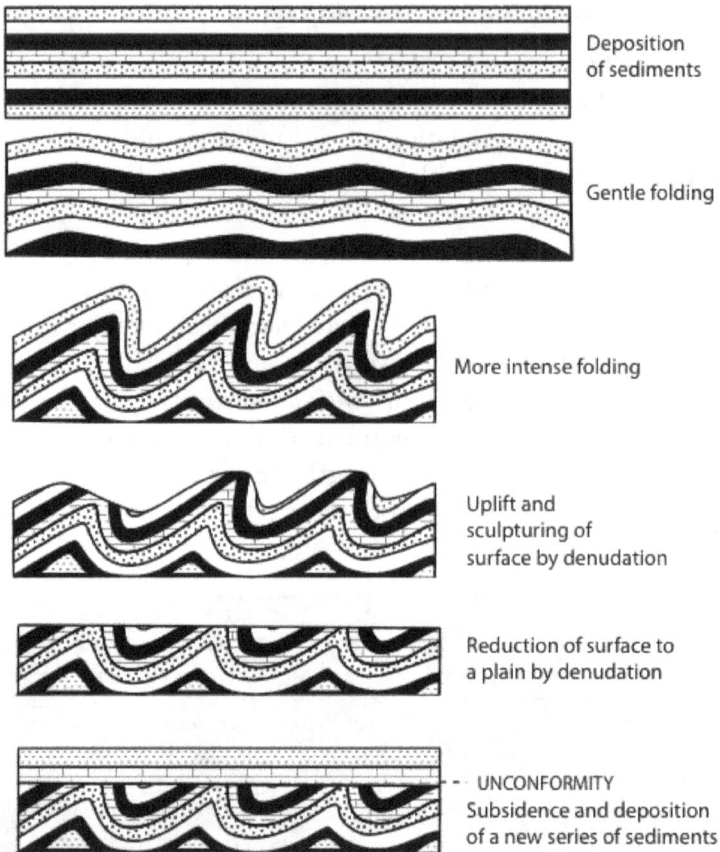

Figure 1.45. Sequence of events in the formation of an unconformity (modified and redrawn from Holmes, 1965).

provides an illustration of unconformable relationships and the sequence of events involved in the generation of an angular unconformity is shown in Fig. 1.45. Examples of angular unconformities are shown in Figs. 1.46 and 1.47.

The sequence exposed in the Grand Canyon, Arizona, provides an excellent case study. Figure 1.48 shows the exposed rock units and Fig. 1.49 places them on a stratigraphic column, indicating the time intervals for which there is no rock record. Several gaps are evident, some of which correspond to very long time intervals; the

Figure 1.46. A strong angular unconformity (http://whaton.uwaterloo.ca/pic/Fig%202,%20Barry.jpg).

Figure 1.47. An angular unconformity (http://written-in-stone-seen-through-my-lens.blogspot.co.uk/2012/01/great-unconformity-of-grand-canyon-part.html).

unconformity between the Cambrian and Devonian represents a gap of more than 70 million years.

In the sub-surface, unconformities are often readily identifiable in seismic sections. Rocks separated by unconformities are commonly

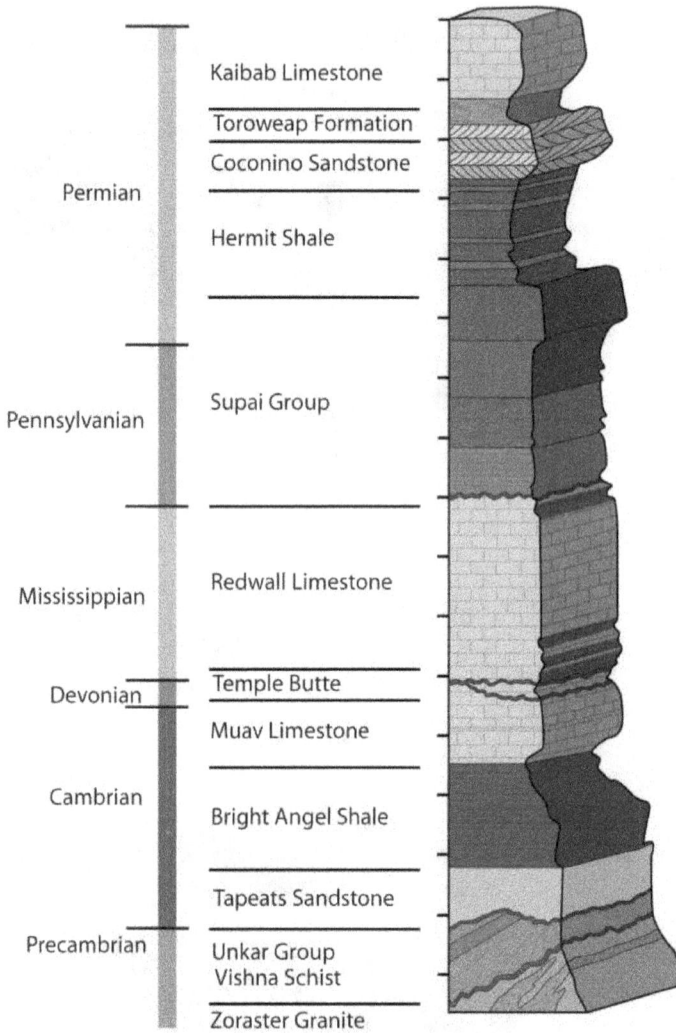

Figure 1.48. Rock sequence exposed at the Grand Canyon (after Marshak, 2005).

characterised by contrasts in their acoustic properties, which enable the surface of separation to act as a good reflector of seismic energy, as shown in Fig. 1.50. Under certain conditions, unconformities can provide trapping opportunities for oil and gas accumulations, as illustrated diagrammatically in Fig. 1.51.

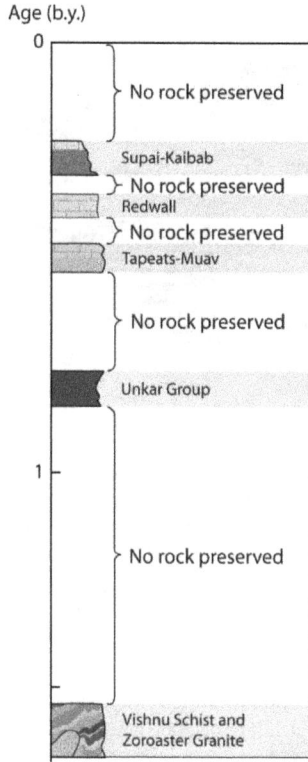

Figure 1.49. Grand Canyon stratigraphic column. Due to unconformities, the strata exposed represent an incomplete record of its geological history (after Marshak, 2005).

### 1.12.2. *Sequence stratigraphy*

Sequence stratigraphy is the study of how sediments within successions vary as the result of sea level changes, which often occur in repeated cycles. It is an important modern approach that allows the prediction of the development and lateral extent of rock types within a sedimentary succession by constructing models of deposition in terms of rises and falls in sea level. The models therefore provide a framework for understanding the distribution of different sediment types in time and space and can be on regional/basin or local scales. Figure 1.52 illustrates the principle of the approach.

A SUBSURFACE UNCONFORMITY

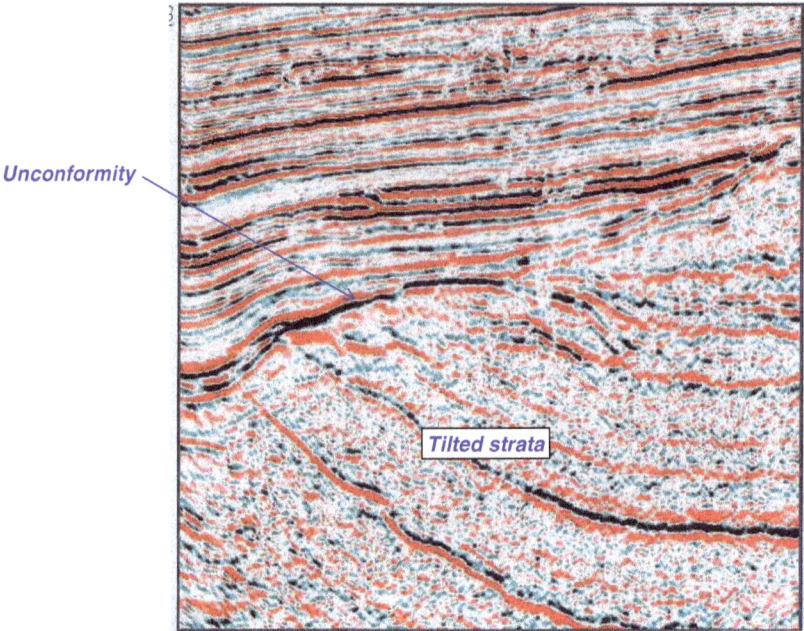

Figure 1.50. Seismic section, offshore Norway, showing an unconformity truncating tilted strata (from GEO EXPro, 2008).

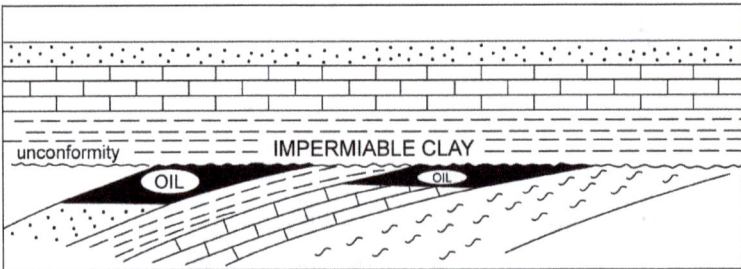

Figure 1.51. Diagrammatic illustration of trapping possibilities associated with an unconformity (modified and redrawn from Fox, 1964).

At any one time, different sediment types are deposited in an area. Shallow water sediments, often coarse clastics, accumulate near land, finer-grained sands and silts further offshore, and clays in deeper water. A transgression occurs when sea level rises, leading to

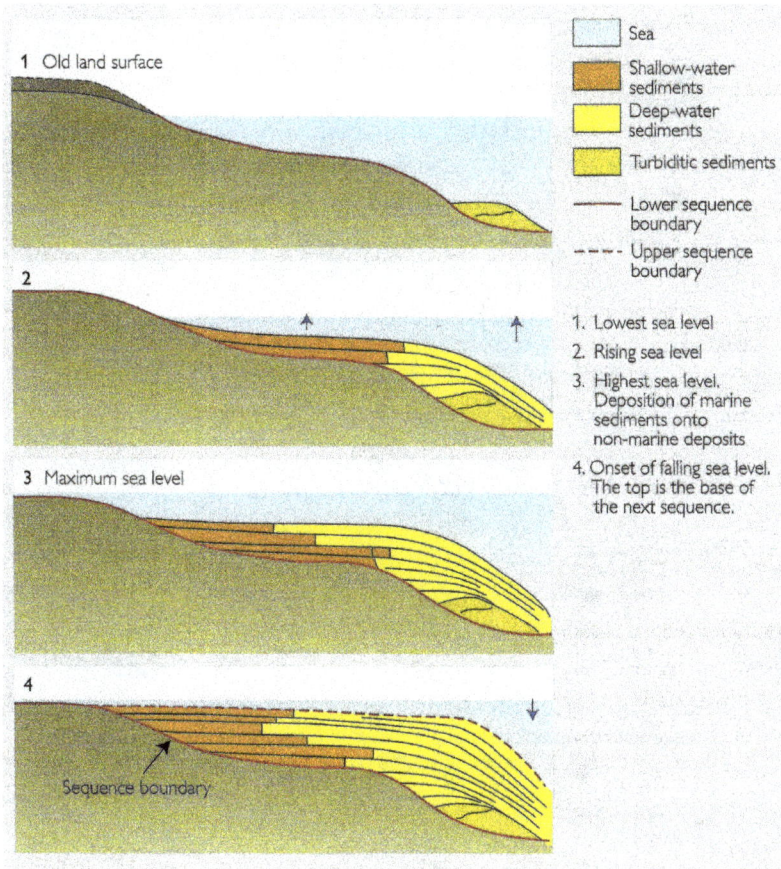

Figure 1.52. Illustration of deposition of a sequence (after Henriksen, 2008).

an increase in the depth of water at any given place and the transition zone from shallow to deep water shifts landwards. A regression follows a fall in sea level, resulting in a decrease in the depth of water and migration of the shallow to deep transition zone towards the sea.

Stratigraphic sequences are packages of sediment associated with these changes in sea level and are composed of a relatively conformable succession of genetically related strata bounded above and below by regional unconformities, as shown in Fig. 1.53. The

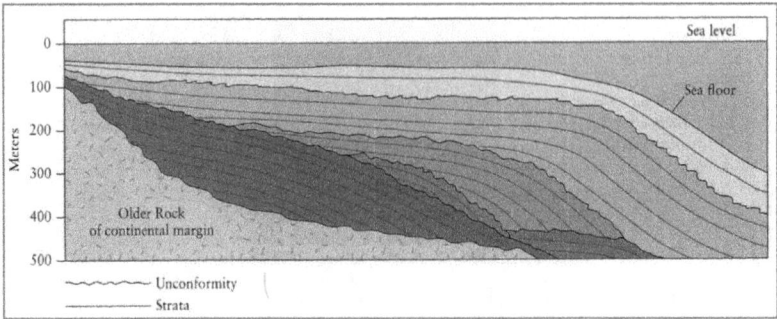

Diagram of a group of sequences bounded above and below by unconformities.
Within each sequence the strata are conformable (after Murck & Skinner, 1999).

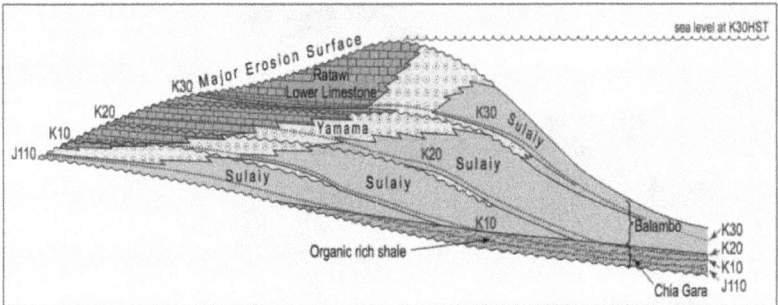

An example of an actual stratigraphic sequence from the Late
Jurassic–Early Cretaceous of southern Iraq (after Aqrawi *et al.*, 2010).

Figure 1.53. Illustration of sequence stratigraphy.

unconformities result from the exposure of the succession following
a regression.

The deposition of an idealised sequence is illustrated in Fig. 1.54.

Although regional/basin-scale packages may exhibit great variations in lithologies in different parts of the basin, they may belong
to the same cycle of sea-level rise and fall. Sequence stratigraphy has
its own extensive terminology and hierarchical stratal elements —
sequences, systems tracts, bounding surfaces, parasequences, to
mention a few.

Integration of information from sequence stratigraphic studies
provides an insight into the history of global changes in sea level
during the Phanerozoic Eon. Although this interpretation remains
controversial, it indicates that the present global sea level is relatively

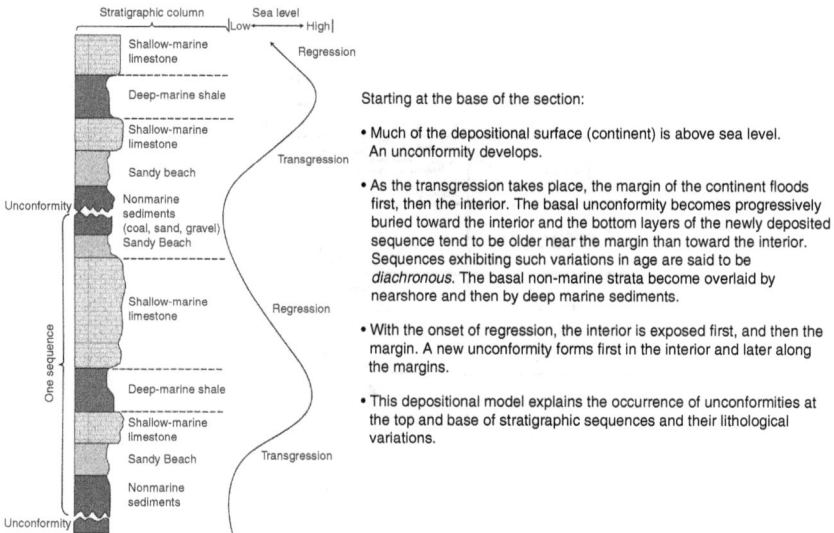

Figure 1.54. The deposition of an idealised sequence (after Marshak, 2005).

low; there appear to have been much higher sea levels in the past, particularly in the Cretaceous and Lower Palaeozoic times (Fig. 1.55).

## 1.13. Correlation

*Correlation* is the demonstration of equivalency. In practice, it involves the identification of certain characteristics in one well and recognition of the same features in a second well or subsequent wells. The intervals exhibiting the same features are considered to be equivalent and can be correlated.

A variety of geological characteristics can be used for correlation purposes but the most widely used are biostratigraphic (fossil content) and lithostratigraphic (rock type) attributes. Surface outcrops can also be correlated.

Figure 1.56 shows a diagrammatic example of a biostratigraphic correlation based on similarities between the fossil contents of a number of beds exposed at three localities. At locality 3, bed D is missing and E directly overlies C. An unconformity is therefore inferred between C and E at locality 3.

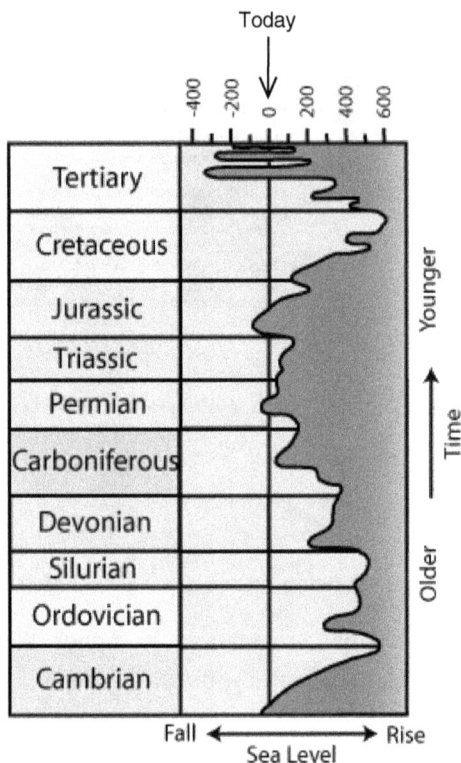

Figure 1.55. The global sea level rise and fall curve during the Phanerozoic Eon (after Marshak, 2005).

Correlation can be on a regional or basin scale, or local, e.g. on a field scale. A diagrammatic example of a basin-scale lithostratigraphic correlation is presented in Fig. 1.57. Unconformities are generally more common towards the basin edge and the section thins in the same direction. A lithostratigraphic correlation panel based on well data, extending regionally across central Iraq, is presented in Fig. 1.58.

An example of a detailed field-scale correlation is shown in Fig. 1.59, demonstrating lateral variations in the thickness and facies of the reservoir in the Scott field, North Sea. Several 2D correlation panels can be combined to produce a 3D picture known as a *fence diagram* (Fig. 1.60).

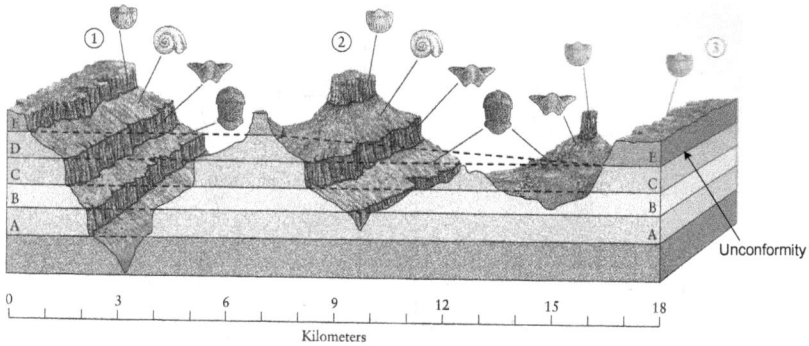

Figure 1.56. Illustration of a biostratigraphic correlation. Correlation is based on similarities between the fossil contents of beds A, B, C, D, and E at three different localities. Bed D is missing at locality 3 and E directly overlies C. An unconformity is therefore present between C and E at locality 3 (after Murck & Skinner, 1999).

Correlation is an important feature of field studies. Its purpose is to identify *flow units* (permeable layers that allow the fluids to flow) in the reservoir and trace their lateral continuity and thickness in the sub-surface. This is illustrated in Fig. 1.61.

## 1.14. Principles of Structural Geology

*Structural geology* is concerned with the study of the deformation of rocks. Manifestations of deformation range from tilting and arching to large- and small-scale fractures and dislocations. The study of the history and processes causing the observed structures is called *tectonics.*

Deductions regarding the deformation experienced by the rocks are based on the present attitude of the bedding planes. By measuring the inclination (dip) of the bedding planes and the trend (strike) of the beds it is possible to work out the structural configuration of the rocks in the area under study. Figure 1.62 provides an illustration of dip and strike and how they are measured, and the procedure for recording this information on maps.

Geological structures include folds, faults and thrusts and these are described briefly below.

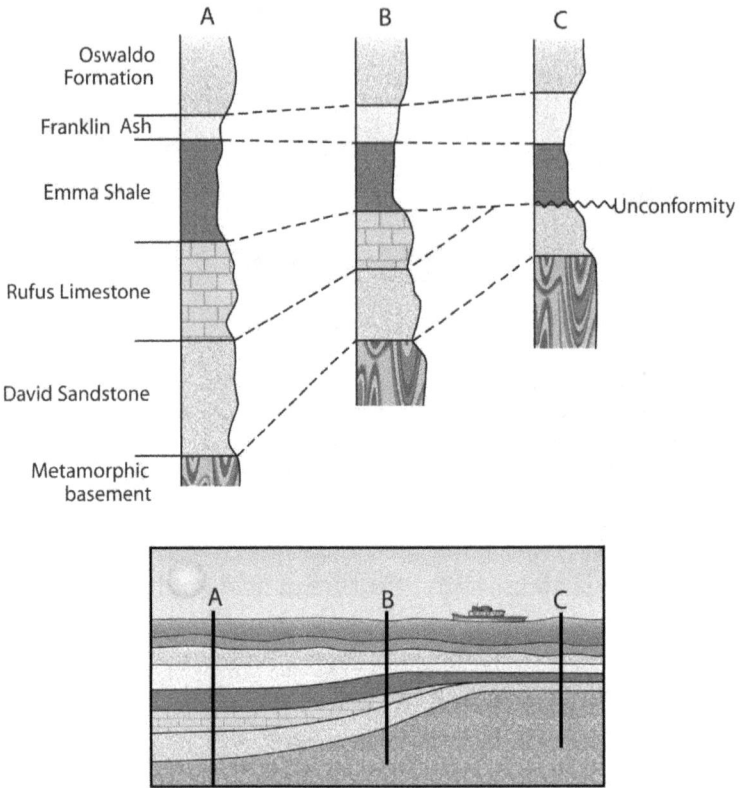

Figure 1.57. Example of a basin scale lithostratigraphic correlation (after Marshak, 2005).

### 1.14.1.  *Folds*

Folds owe their origin to crumpling or arching of strata.

#### 1.14.1.1.  *Anticline*

An *anticline* is an arch in which the two sides, usually called limbs or flanks, dip away from one another. The strata forming the centre of the fold are referred as the "core" and those forming the outer part are called the "envelope". In an anticline the core is older than the envelope, as depicted in Fig. 1.63. Field photographs of anticlines are presented in Figs. 1.64 and 1.65.

Anticlines are the most common oil and gas traps and a typical anticlinal trap is shown in Fig. 1.66.

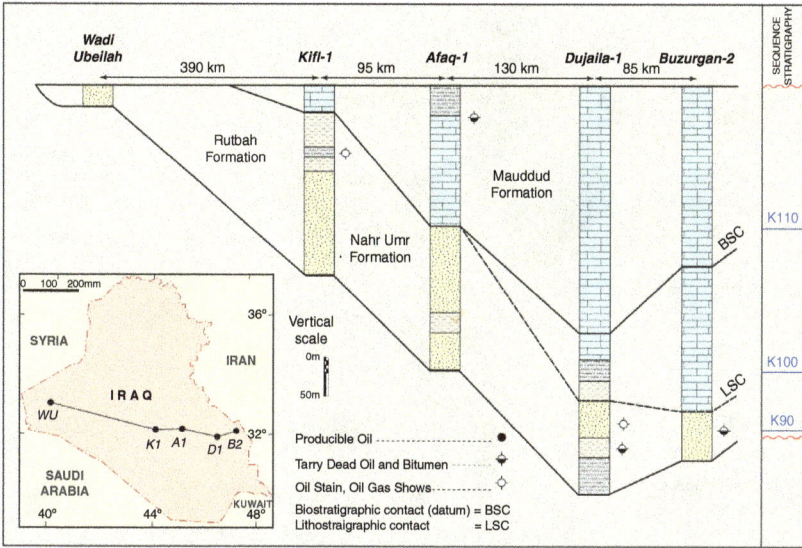

Figure 1.58. Regional east–west lithostratigraphic correlation panel based on well data, central Iraq. Note the marked eastward thickening of the carbonate sequence (after Agrawi *et al.*, 2010).

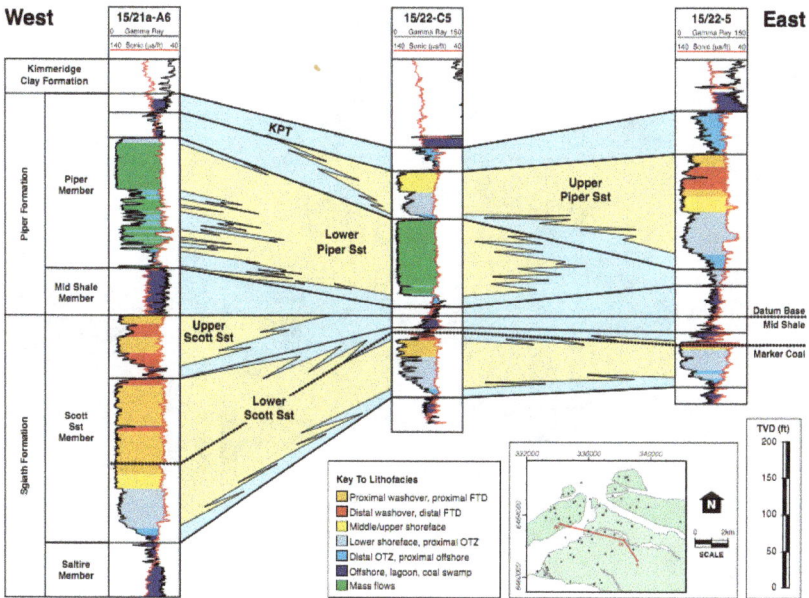

Figure 1.59. East–west correlation section through the Scott field, North Sea, showing lateral reservoir thickness and facies variations in the Scott field, North Sea (after Guscott *et al.*, 2003).

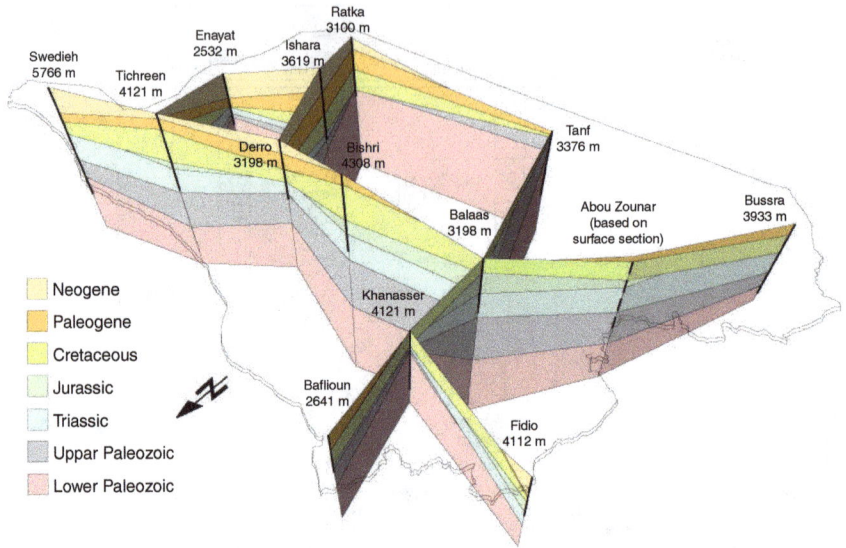

Figure 1.60.  3D presentation of correlation panels. Several 2D correlation panels can be combined to produce a 3D picture known as a *fence diagram* (after Brew, 2001).

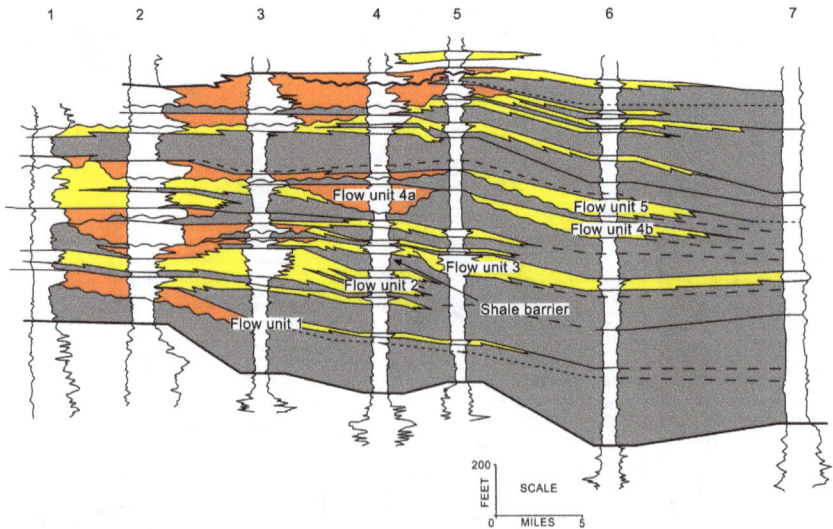

Figure 1.61.  Correlation panel showing the complexity of field scale distribution of flow units (modified and redrawn from Bhattachrya, 2006).

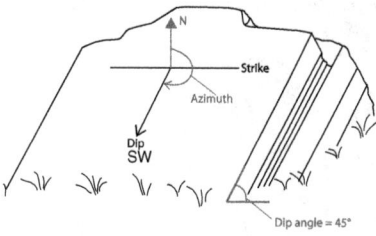

- *Dip* and *strike* are two fundamental structural properties of layered rocks.

- Dip is the maximum amount of inclination from the horizontal.

- Strike indicates the trend of the beds and is perpendicular to dip.

- *Azimuth* is dip direction, expressed as a clockwise rotation from the **north** and should not be confused with dip.

- On maps, unless otherwise in dicated, north is up the page.

- In this example, dip = 45°, azimuth is southwesterly, about 225°, and strike approximately NW–SE. On a map this would be recorded as

45°

- Measuring dip and strike and plotting them on a base map allows the structure of the area under study to be interpreted.

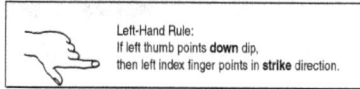

Left-Hand Rule:
If left thumb points **down** dip,
then left index finger points in **strike** direction.

Figure 1.62. Illustration of dip and strike and their measurement.

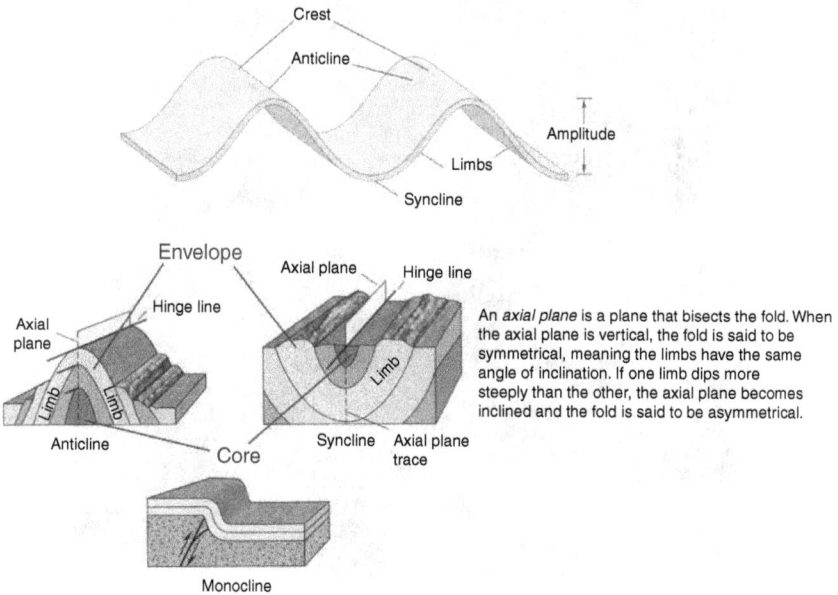

An *axial plane* is a plane that bisects the fold. When the axial plane is vertical, the fold is said to be symmetrical, meaning the limbs have the same angle of inclination. If one limb dips more steeply than the other, the axial plane becomes inclined and the fold is said to be asymmetrical.

Figure 1.63. Diagrammatic illustration of folds and associated terminology (after Marshak, 2005).

Figure 1.64. A small-scale, low-amplitude, symmetrical anticline (after Marshak, 2005).

Figure 1.65. A large, high amplitude, symmetrical anticline (after Hull & Warman, 1970).

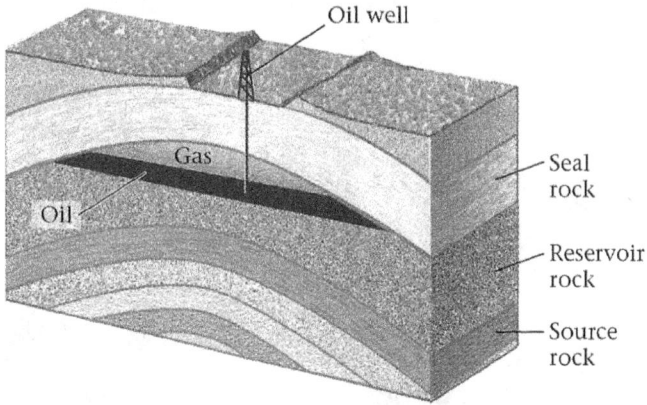

Figure 1.66. A typical anticlinal trap (after Marshak, 2005).

Figure 1.67. An asymmetrical syncline (photo by J. Cosgrove).

### 1.14.1.2.  *Syncline*

A *syncline* is a fold in which the limbs dip towards one another. In the case of a syncline, the core is younger than the envelope (Fig. 1.63). A field photograph of a syncline is shown in Fig. 1.67.

Figure 1.68. Asymmetrical folds — axial planes inclined (modified from Marshak, 2005).

Figures 1.64 and 1.65 depict "symmetrical" folds, characterised by vertical axial planes, defined as planes that bisect the folds. If one limb dips more steeply than the other, the axial plane becomes inclined and the fold is said to be "asymmetrical". Examples of asymmetrical folds are presented in Figs. 1.68–1.70.

### 1.14.1.3. *Monocline*

A *monocline* is a simple step-like flexure in which more or less horizontal beds locally assume a dip in one direction and then flatten out again. A field photograph of a monocline is shown in Fig. 1.71.

### 1.14.2. *Faults*

A *fault* is a fracture or dislocation in the Earth's crust along which there has been displacement of the rocks on one side relative to those on the other. The surface on which movement takes place is referred to as the *fault plane*. With few exceptions, fault planes are normally inclined, which makes it possible to distinguish between the rocks lying above the fault plane and those below it.

Figure 1.69. Strongly asymmetrical anticline (photo by M. Ala).

Figure 1.70. Small-scale asymmetrical structures, Stair Hole, Dorset, SW England (photo by M. Ala).

Figure 1.71. A monocline (http://coloradogeologicalsurvey.org/colorado-geology/structures/folds/).

The former are called the *hanging wall* and the latter are known as the *footwall*. Faults are classified in accordance with the direction of the displacement associated with them and the crustal stress regime responsible for their development. Figure 1.72 illustrates the various fault types, the mechanisms of their formation and the associated terminology.

If the hanging wall slips down with respect to the footwall, the fault is categorised as *normal* and is formed by tensional forces. Normal faulting extends the crust and causes subsidence. Other features associated with normal faulting include *horsts* and *grabens*. A horst is an upfaulted block while a graben is a downfaulted block. Sometimes the blocks undergo rotation, forming *tilted fault blocks* and *half-grabens* (Fig. 1.72).

When the footwall slips up compared to the hanging wall, the fault is classified as *reverse* and results from compressional forces. Reverse faulting shortens the crust and causes uplift.

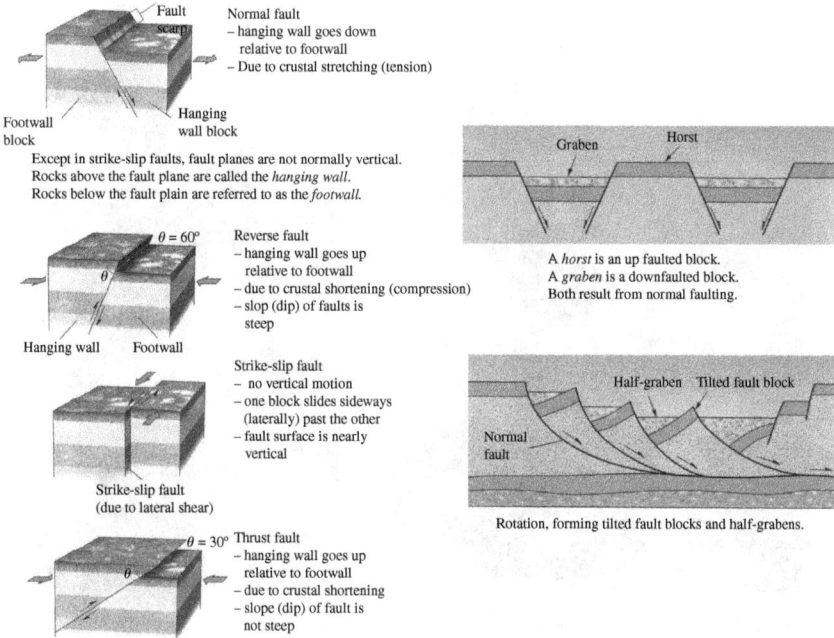

Figure 1.72. Fault types, associated structures and terminology (after Marshak, 2001, 2005).

Field photographs showing normal and reverse faults are presented in Figs. 1.73 and 1.74, respectively.

Features associated with normal and reverse faults can form oil and gas traps as shown diagrammatically in Fig. 1.75. The Gulf of Suez serves as an example of traps associated with horsts and tilted fault blocks (Fig. 1.76).

For the sake of completeness, mention should be made of *listric* faults, which are normal faults characterised by curved fault planes, as illustrated in Fig. 1.77. The dip decreases with depth but the throw of the fault increases with depth, the movement being contemporaneous with sedimentation; i.e. the fault "grows" (and hence features of this type are also known as *growth faults*) as the sediments are being deposited. Consequently, the strata thicken on the downthrown side of the fault. They also roll over into the fault plane, forming anticlines, the axes of which shift with depth, resulting in the crests of successive anticlines not being vertically aligned.

Figure 1.73. A field photograph of a normal fault. (http://www.appstate.edu/~marshallst/photos/mass_photos/utah/utah-056.jpg).

This structural style develops in thick delta sequences and accumulations in the prolific Niger delta are associated with roll-over traps of this type.

In the case of *strike-slip faults*, the displacement is primarily horizontal with little or no vertical component and the fault plane is normally vertical (Fig. 1.72).

### 1.14.3.   *Thrusts*

A *thrust* is a low-angle reverse fault along which there has been a great deal of movement. Thrust-associated oil and gas traps are known from several parts of the world and one such example, the Turner Valley field, western Canada, is presented in Fig. 1.78. Using the horizontal scale on the cross-section, the magnitude of movement can be estimated; it exceeds a mile.

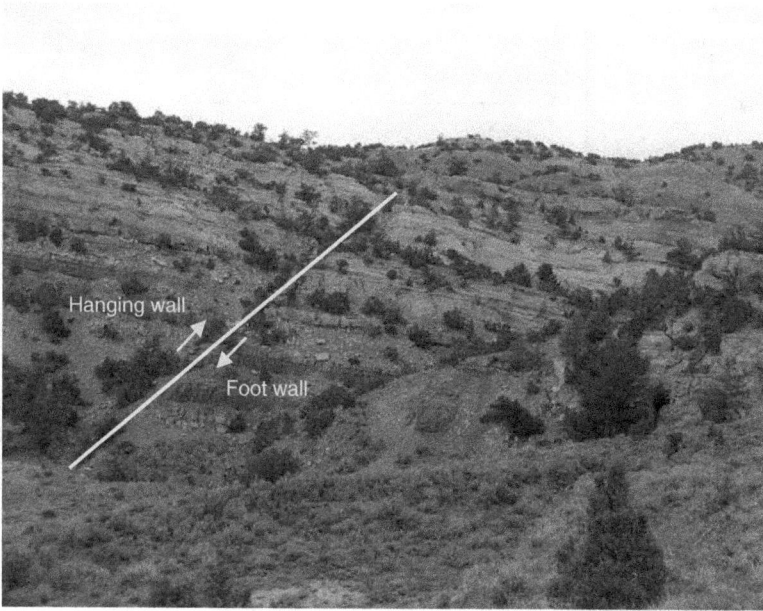

Figure 1.74. A field photograph of a reverse fault (http://fieldcamp.missouri. edu/images/P6100121_Dallas%20Dome%20Faulting.jpg).

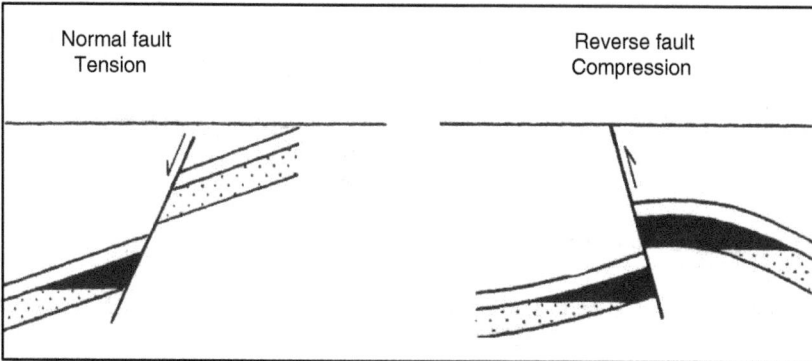

Figure 1.75. Traps associated with faults. Both normal and reverse faults can provide traps. The essential requirement is that the permeable bed (reservoir) must be juxtaposed across the fault against an impervious layer to prevent the hydrocarbons from further movement (after Stoneley, 1995).

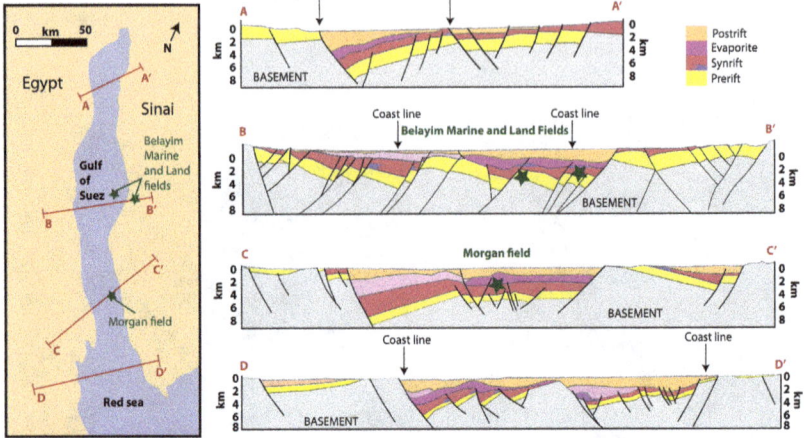

Figure 1.76. Gulf of Suez. Example of oil fields associated with horsts and tilted fault blocks (modified and redrawn from http://gsabulletin.gsapubs.org/content/112/12/1877/F3.large.jpg).

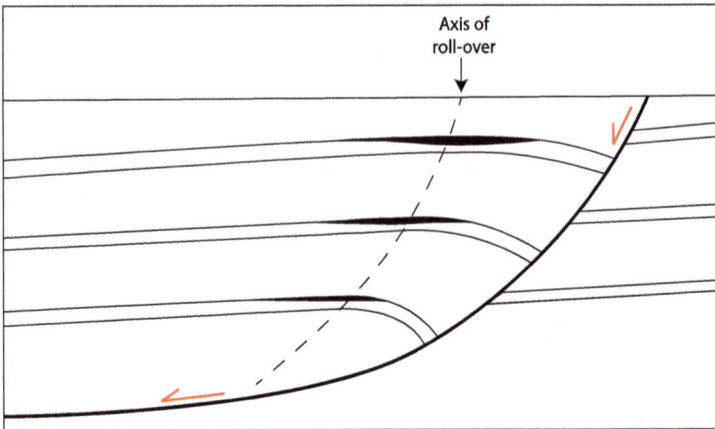

Figure 1.77. Listric fault associated roll over anticlines (after Stoneley, 1995).

### 1.14.4.  *Structures formed by the movement of evaporitic rocks*

Evaporitic rocks exhibit a particular characteristic: They behave in a plastic manner and flow when subjected to stress. The flow takes place from regions of high stress to those of low stress, causing

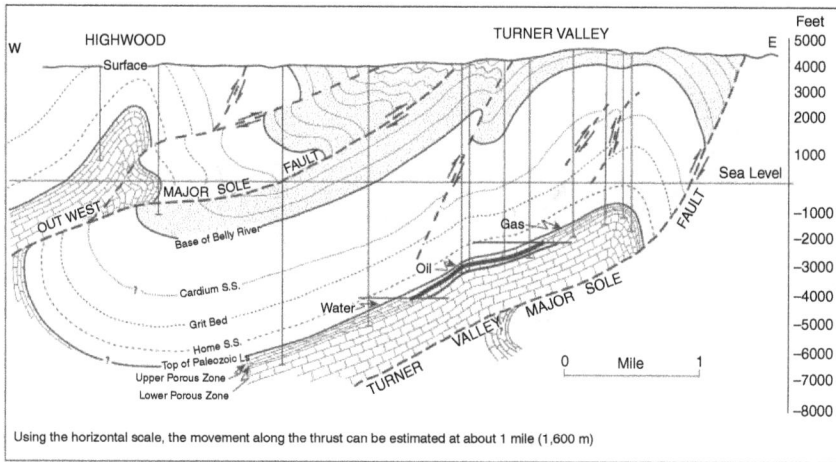

Figure 1.78. Cross-section through the Turner Valley oil field, western Canada. An example of a thrust-associated trap (modified and redrawn from Levorsen, 1967).

great variations in the thicknesses of the layer over short distances. Layers exhibiting this property are called "incompetent" rocks, as opposed to "competent" units, which respond to stress by buckling and faulting without changes in their shape or thickness. Sandstones, limestones and dolomites fall in the latter category.

Once triggered, flowage gives rise to salt-induced features, the initial stage of which is marked by the formation of a bulge or salt pillow. Provided there is sufficient supply of the evaporitic material, the pillow will grow and penetrate the overlying strata and reach the surface in some cases. These features are known as piercement domes, plugs or diapirs and sometimes a number of adjacent plugs coalesce and form salt walls. Figure 1.79 provides a diagrammatic illustration of the various types of salt-induced structures. Collectively, the processes involved in the generation of these features are referred to as salt tectonics or diapirism.

Another factor that aids the rise of salt is the contrast between its density and that of the overlying rocks. Salt has a density of $2.03 \, \text{g/cm}^3$, which is significantly lower than that of the other common sedimentary rocks such as sandstone (comprised mostly

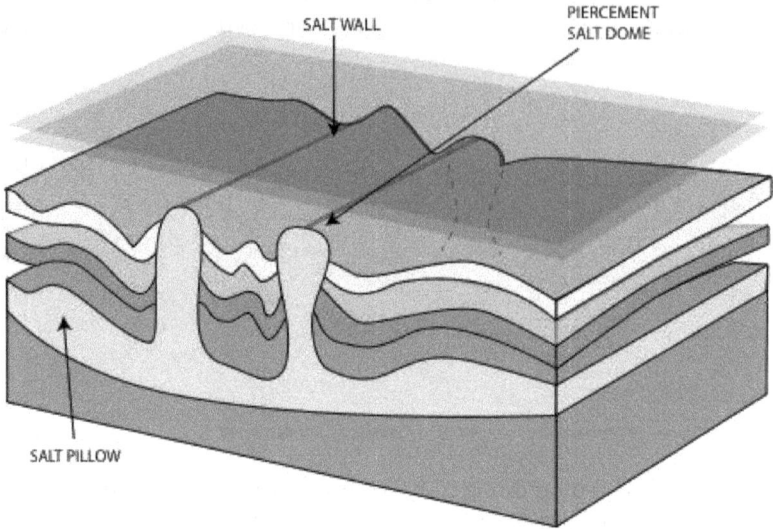

Figure 1.79. Diagrammatic illustration of salt-induced features (modified and redrawn from former UKOOA).

of quartz, density $2.65\,\mathrm{g/cm^3}$), limestone (made up of calcite, density $2.71\,\mathrm{g/cm^3}$) and dolomite (consisting of dolomite, density $2.87\,\mathrm{g/cm^3}$). The buoyancy force created by this density contrast enables the salt to rise through the overlying blanket of sediments.

Salt-induced structures produce a large variety of oil and gas traps (see Chapter 4).

## References

Aqrawi, A.M.A., Goff, J.C., Horbury, A.D. *et al.* (2010). *The Petroleum Geology of Iraq,* Scientific Press, UK.

Bhattacharya, J.P. (2006). Applying Deltaic and Shallow Marine Outcrop Analogs to the Subsurface, Search and Discovery Article #40192.

Brew, G. (2001). Tectonic evolution of Syria interpreted from integrated geophysical and geological analysis, unpublished PhD dissertation, Cornell University, Ithaca, NY.

Cochise College Virtual Geology Museum (2011). Available online at: http://skywalker.cochise.edu/wellerr/rocks/sdrx/limestone13.htm.

Flint, R.F. and Skinner, B.J. (1974). *Physical Geology,* John Wiley, New York.

GEO EXPro (2008). *A Minute to Read*, **5**(4), 14–18.

Fox, A.F. (1964). *The World of Oil*, Pergamon Press, Oxford.

Guscott, S., Russell, K., Thickpenny, A. and Poddubiuk, R. (2003). "The Scott Field, Blocks 15/21a, 15/22, UK North Sea", in Gluyas, J.G. and Hichens, H.M. (eds.), *United Kingdom Oil and Gas Fields Commemorative Millennium Volume Geological Society Memoir 20*, The Geological Society Publishing House, Bath, England.

Harbaugh, J.H. (1965). "Carbonate Reservoir Rocks", in Chillingar, G.V., Bissell, H.J. and Fairbridge, R.W. (eds.), *Carbonate Rocks*, Elsevier Publishing Company, London, UK.

Henriksen, N. (2008). *Geological History of Greenland*, Geological Survey of Denmark and Greenland (GEUS), Copenhagen, Denmark.

Holmes, A. (1965). *Principles of Physical Geology*, 2$^{nd}$ Edition, Nelson & Sons.

Hull, C.E. and Warman, H.R. (1970). "Asmari Oil Fields of Iran", in Halbouty, M.T. (ed.), *Geology of Giant Petroleum Fields*, AAPG, Tulsa, OK.

Krumbein, W.C. and Sloss, L.L. (1963). *Stratigraphy and Sedimentation*, W.H. Freeman, San Francisco.

Levorsen, A.I. (1967). *Geology of Petroleum*, 2$^{nd}$ Edition, W.H. Freeman, San Francisco.

Marshak, S. (2001). *Earth: Portrait of a Planet*, W.W. Norton, New York, NY.

Marshak, S. (2005). *Earth: Portrait of a Planet*, 2$^{nd}$ Edition, W.W. Norton, New York, NY.

Murck, B.W. and Skinner, B.J. (1999). *Geology Today: Understanding Our Planet*, John Wiley & Sons, New York, NY.

Schofield, N. (2015). Volcanic Rocks and the Petroleum System West of Shetlands (WoS), *PESGB Magazine*, **January**, 7–9.

Sellwood, B.W. (2007). *Terrigenous Clastic Reservoir Rocks*, Department of Earth Science and Engineering, Imperial College London, UK.

Stoneley, R. (1995). *An Introduction to Petroleum Geology for Non-Geologists*, Oxford University Press, Oxford, UK.

The Rock Cycle (1992). Available online at: http://www.manitoba.ca/iem/min-ed/kidsrock/origins/images/rockcycle.png.

UKOOA. United Kingdom Offshore Oil Operators Association, now Oil & Gas UK.

# Chapter 2

# Controls on Oil and Gas Occurrence:
# Sedimentary Basins and Plate Tectonics

## 2.1. Factors that Determine Hydrocarbon Formation

Oil and gas fields are commercially significant concentrations of hydrocarbons. Every oil and gas accumulation is the result of the following fundamental processes:

- generation,
- migration,
- accumulation.

Generation takes place in *source rocks*. The hydrocarbons then *migrate* from the source beds and *accumulate* in permeable formations known as *reservoir rocks*. Accumulation requires the presence of a *trap* in the reservoir and, in order to prevent further upward hydrocarbon movement, the reservoir must be covered by an impermeable cover called a *seal* or *cap rock*.

Source, reservoir, trap and seal are the parameters that control the occurrence of oil and gas accumulations (fields). These parameters are geologically controlled.

## 2.2. Geological Controls on Oil and Gas Occurrence

### 2.2.1. *Source rocks*

A *source rock* is a fine-grained sediment that in its natural setting has generated and released sufficient hydrocarbons to form a commercial accumulation of oil and gas. Source rocks are clay or carbonate organic-rich muds deposited under low-energy, anoxic conditions. To function as a source rock, the sediment must contain a minimum of 2% by weight of organic matter.

The single most important factor in the generation of hydrocarbons in a source rock is temperature. The action of heat on the organic matter (kerogen) contained in the sediment leads to the formation of oil and gas. For temperature to rise to the required level, the source rock must be buried and attain what is known as the minimum level of *maturity.*

### 2.2.2.  *Reservoir rocks*

A *petroleum reservoir* is a permeable rock in communication with a mature source bed. Sandstones and carbonates form the overwhelming majority of reservoirs worldwide. Under special circumstances, igneous and metamorphic rocks can also act as petroleum reservoirs.

### 2.2.3.  *Traps*

A *trap* is a special situation in the reservoir that arrests the migration process and causes the hydrocarbons to accumulate. Traps may be formed by changes in the shape of the reservoir (structural traps), by variations in its lateral continuity (stratigraphic traps) or by a combination of these factors (combination traps). Trap formation must predate hydrocarbon migration.

### 2.2.4.  *Seals or cap rocks*

*Seals* are impervious beds that overlie reservoirs and prevent further upward movement. Generally, this prevents the loss of hydrocarbons. Evaporites and shales are the most common seals.

## 2.3.  Sedimentary Basins

### 2.3.1.  *Definition and origin*

Worldwide, oil and gas fields are associated with *sedimentary basins,* the distribution of which is shown in Fig. 2.1.

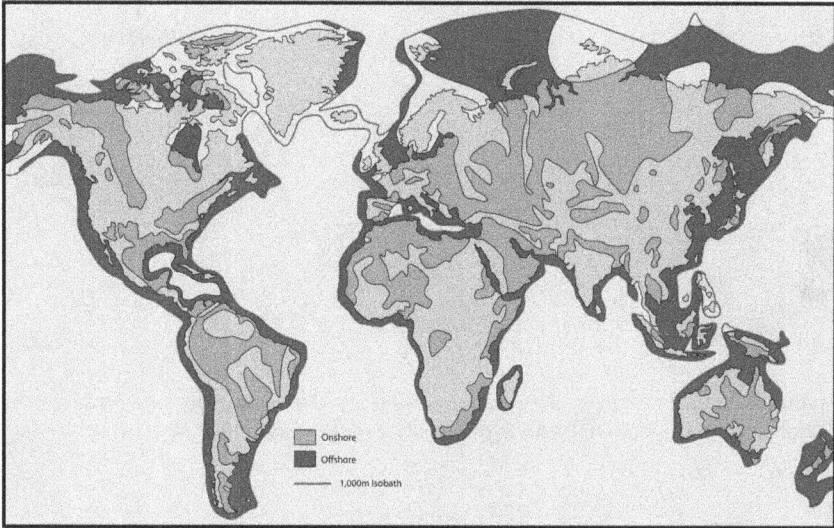

Figure 2.1. Global distribution of the major sedimentary basins (modified and redrawn from Kirby, 1977).

A sedimentary basin is a depression in the Earth's crust containing a large thickness of sedimentary rocks. Generally, their formation results from the thinning and stretching of the *lithosphere* or crust, caused by an upwelling of the underlying *asthenosphere.* Stretching causes tension in the crust, which leads to normal faulting (*rifting*) and subsidence. A present day analogue for this stage of basin development is the East African Rift system, as depicted in Fig. 2.2.

At the same time, the rising, hot asthenosphere heats up the thinned lithosphere. When the rifting stops, the lithosphere cools, thickens and becomes denser. The heavier lithosphere sinks, causing further subsidence. This sequence of events is depicted in Fig. 2.3. Total subsidence is represented by thermal subsidence + mechanical subsidence (extensional faulting) + sediment loading as illustrated in Fig. 2.4.

The North Sea serves as an excellent basin formation case study. In its present form, it began to develop in the Permian over an asthenospheric upwelling that caused stretching and tension in the overlying lithosphere, as shown in the regional crustal section in

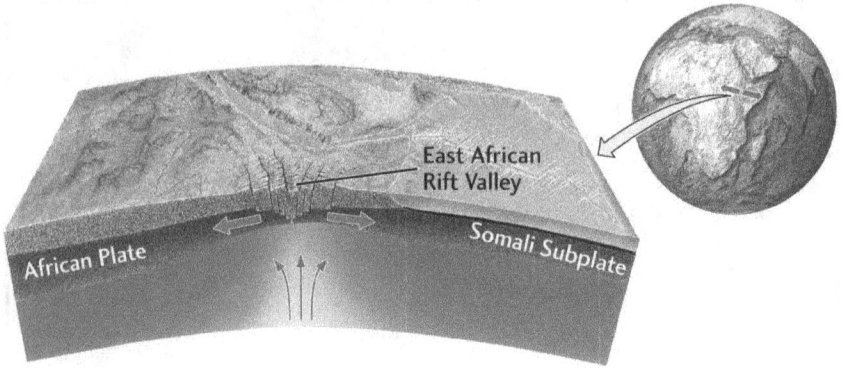

Figure 2.2. East African Rift system. Modern day analogue representing the initial-stage sedimentary basin formation (from Sellwood, 2007).

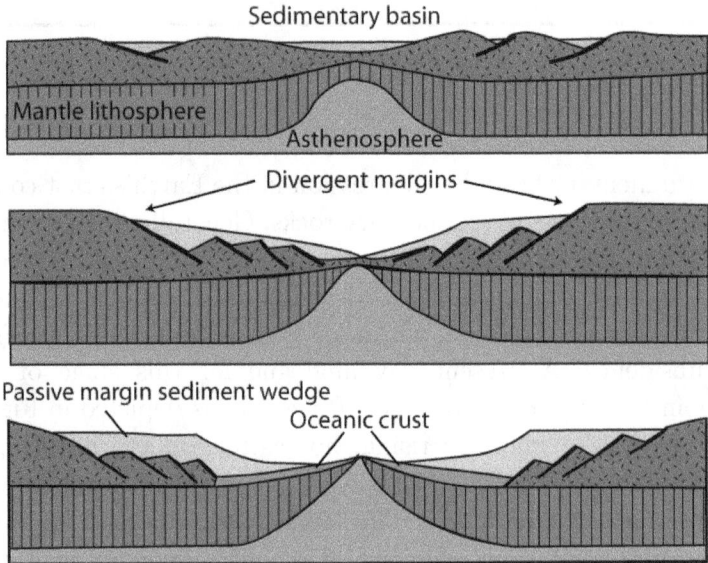

Figure 2.3. Sequence of events in the formation of a sedimentary basin (modified and redrawn from Allen & Allen, 2005).

Fig. 2.5. Normal faulting followed, leading to subsidence, which in turn brought about sedimentation. Strong rifting occurred in the Jurassic, resulting in the formation of a major graben system: the north–south-trending Viking Graben in the northern North Sea

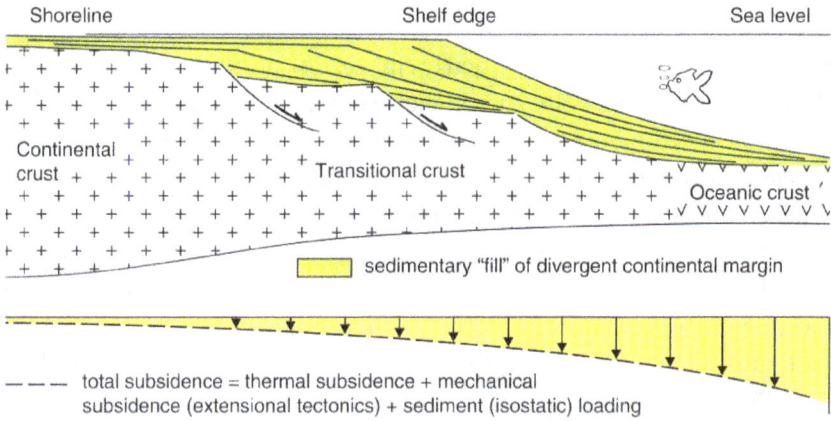

Figure 2.4. Diagrammatic section illustrating subsidence patterns. Subsidence rate increases in the seaward direction (after Catuneanu, 2006).

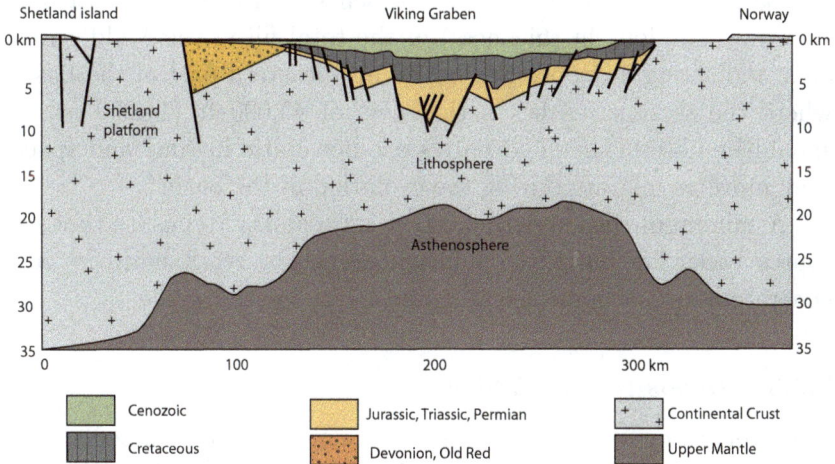

Figure 2.5. East–west crustal section through the northern North sea, showing the Permo-Triassic asthenospheric upwelling that resulted in extension and subsidence throughout the Mesozoic and Tertiary (modified and redrawn from Ziegler, 1990).

and the northwest–southeast-trending Central Graben in the central North Sea. The great majority of the oil fields in the North Sea lie in these grabens and the accumulations are associated with horsts (upfaulted blocks) and tilted fault blocks.

### 2.3.2.   *Characteristics*

Sedimentary basins vary greatly in terms of their aerial extent, shape, thickness and the age of their sedimentary fill. Accumulation of a thick sedimentary succession is dependent on *subsidence* — the sinking of the basin floor over a long period of time (tens of millions of years) — creating *accommodation space*, which in turn leads to sedimentation. Subsidence is therefore a fundamental feature of sedimentary basins. However, it is not uniform throughout the basin; some parts of the basin floor subside more rapidly than others and thus acquire a thicker sedimentary section. This is referred to as *differential subsidence*. The variations in the sedimentary section resulting from differential subsidence in the Middle East Basin are demonstrated in the cross-section in Fig. 2.6.

The area of greatest sediment thickness at any given time is called the basin *depocentre*. Figure 2.7 presents an isopach map (a map showing variations in thickness) of the total fill in the Middle East area, indicating a depocentre in the northeastern part of the basin, where the section reaches a thickness of 45,000 ft (14,000 m). It should be noted that depocentres are not static in time and space; they move or migrate during the evolution of the basin.

A minimum thickness of 3,000 m is necessary to ensure that the source rocks are buried to sufficient depth to reach maturity and generate hydrocarbons (see Chapter 4).

### 2.3.3.   *Depositional history*

Sea-level changes control the accumulation of the sediments that constitute the fill of sedimentary basins. The depositional history of a sedimentary basin is characterised by a series of *transgressions* and *regressions*. A transgression is defined as a relative rise in sea level, resulting in an advance of the sea over the land. There is an increase in the depth of water, leading to the deposition of finer-grained sediments. A regression represents a relative fall in sea level, resulting in a withdrawal of the sea from the land. The depth of water decreases and results in the deposition of generally coarser-grained sediments. These events are illustrated in Fig. 2.8.

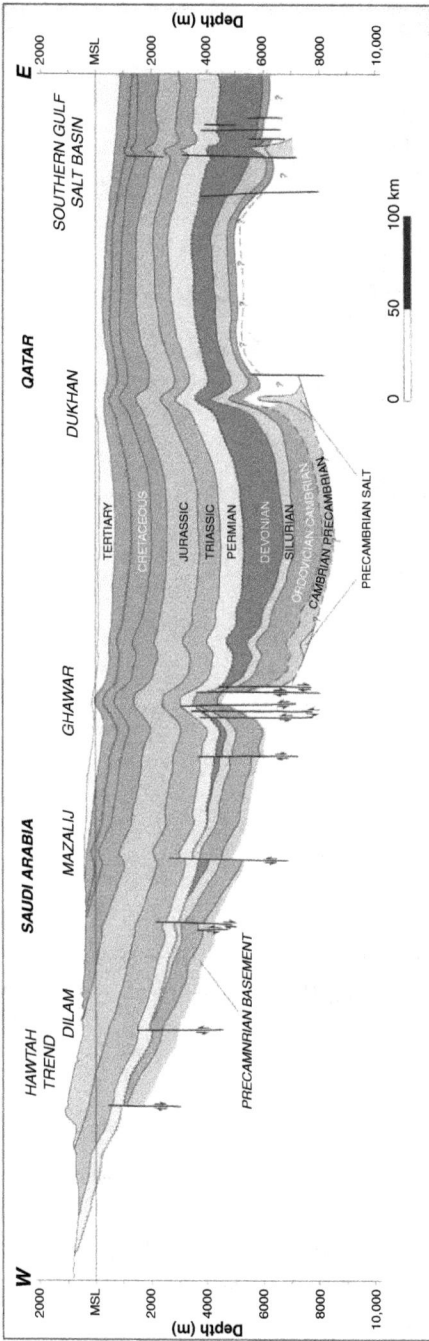

Figure 2.6. East–west cross-section through the Middle East Basin, showing variations in the thickness of the sedimentary section resulting from differential subsidence (after Konert *et al.*, 2001).

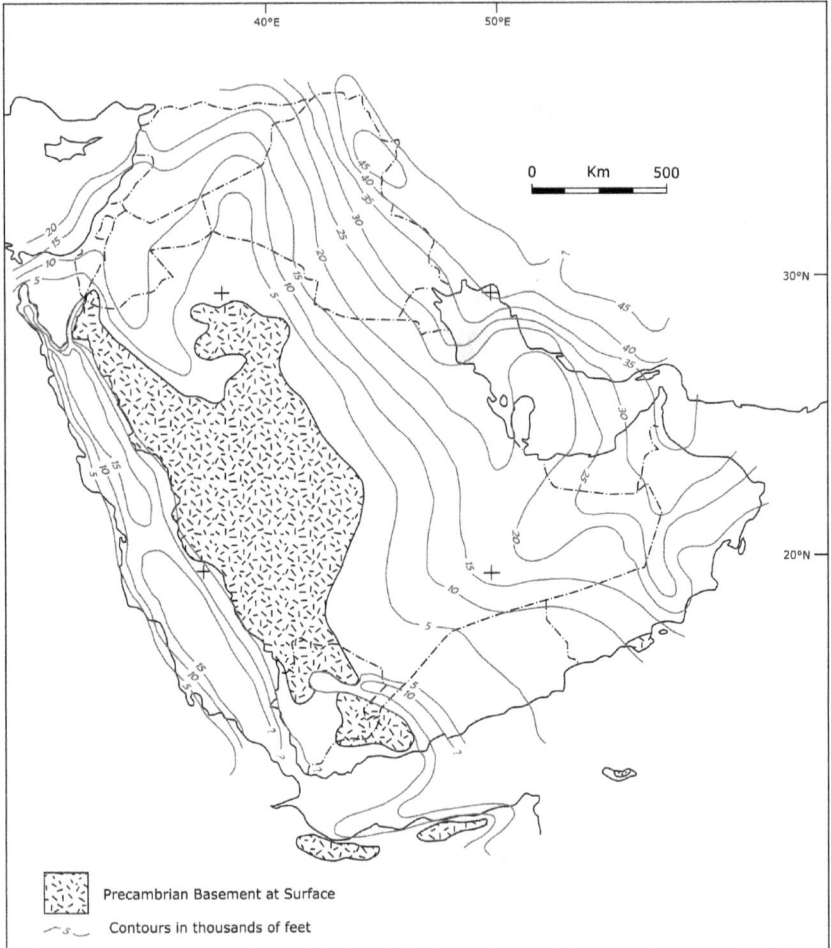

Figure 2.7. Isopach map showing the total sedimentary fill in the Middle East Basin. Note the depocentre in the northeastern part of the basin where the section reaches a thickness of 45,000 ft (modified and redrawn from Beydoun, 1991).

## 2.3.4.  *Sedimentary basins and hydrocarbons*

The association between sedimentary basins and petroleum occurrence is a natural one since the geological conditions necessary for the generation and accumulation of oil and gas are fulfilled in a sedimentary basin. Within the basin fill there are organic-rich sediments that can function as source rocks, permeable rocks that

Transgression ⟶

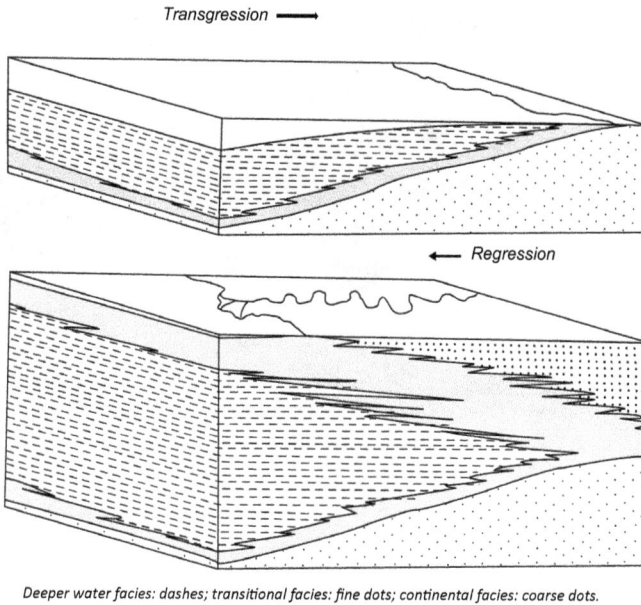

⟵ Regression

*Deeper water facies: dashes; transitional facies: fine dots; continental facies: coarse dots.*

Figure 2.8. Depositional history of sedimentary basins (modified and redrawn from Chapman, 1977).

act as reservoirs and impermeable formations that provide the seals. Traps are formed by the movements affecting the basin fill (tectonic history of the basin) or variations in the lateral continuity of the reservoir beds due to facies changes.

## 2.4. Plate Tectonics and the Formation and Evolution of Sedimentary Basins

*Plate tectonics* is a process that explains and links together a variety of geological and geophysical phenomena. These include:

- mountain building,
- seismicity,
- volcanicity,
- metamorphism,
- rift zones,
- mid-ocean ridges,

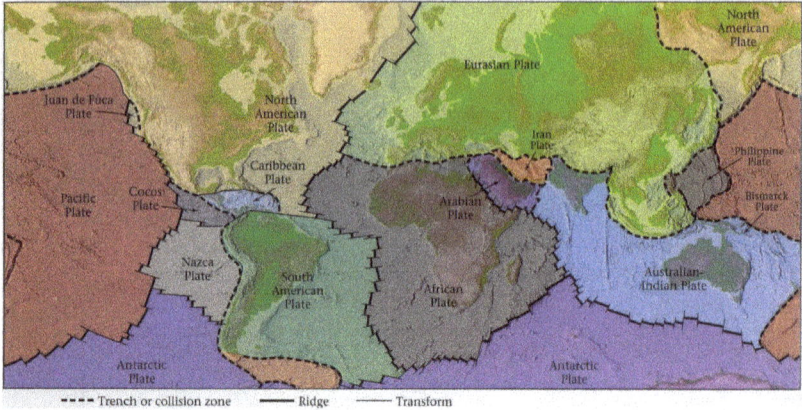

Figure 2.9. The major plates making up the lithosphere (after Marshak, 2005).

- deep sea trenches,
- island arcs,
- continental drift and ocean floor spreading,
- distribution of fossil faunas, floras and other geological features,
- sedimentary basin formation and evolution.

### 2.4.1. *Overview of plate tectonics*

The theory of plate tectonics was introduced into the literature in 1967–1968 and is now a cornerstone of modern Earth Sciences. It rests on the following principles.

- The Earth's crust or lithosphere is divided into a number of internally rigid blocks or plates. The plates vary in size and their thickness ranges from 10 km in the vicinity of mid-ocean ridges to about 40 km over the continents. The major plates making up the lithosphere are outlined in Fig. 2.9.

  The plates move at widely different rates. The most rapid rates, over 18 cm per year, are observed in the Pacific Ocean, while plates underlying the Atlantic and the western parts of the Indian Oceans are characterised by relative velocities of 1.7–4.7 cm per year. Plate movement rates and the directions of travel are shown in Fig. 2.10. Continents are superficial passengers on plates. Where a continent and the adjacent ocean floor ride on the same plate, the ocean

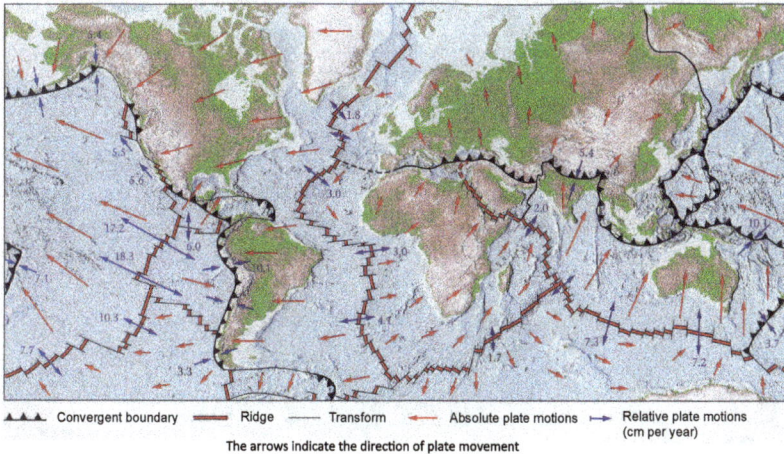

Figure 2.10. Rates of movements of the Earth's principal plates (after Marshak, 2005).

expands with essentially aseismic margins (e.g. the Atlantic, which is growing at the expense of the Pacific). By contrast, where the continental–oceanic crust boundary separates two different plates, one of which is underthrusting the other, a compressive tectonic regime develops. Such a zone is seismically active and is characterised by a deep trench on the oceanward side and a mountain belt along the edge of the continental block (e.g. the Pacific coast of South America).

• The ocean floors are young and the oldest rocks found there are of Jurassic age (150–180 million years). Maps showing the age of the ocean floors are presented in Fig. 2.11. It should be noted that the floor becomes older with increasing distance from the ridges, which are present in all the oceans as shown in Figs. 2.12–2.15.

• The plates are continuously in motion relative to one another and nearly all seismic activity, volcanicity and tectonic deformation take place along plate margins. There are three types of plate margin:

(a) *Accreting, constructive* or *divergent.* At these margins, new crust is created by magmatic accretion or extrusion, resulting in the spreading of the sea floor and the expansion of the

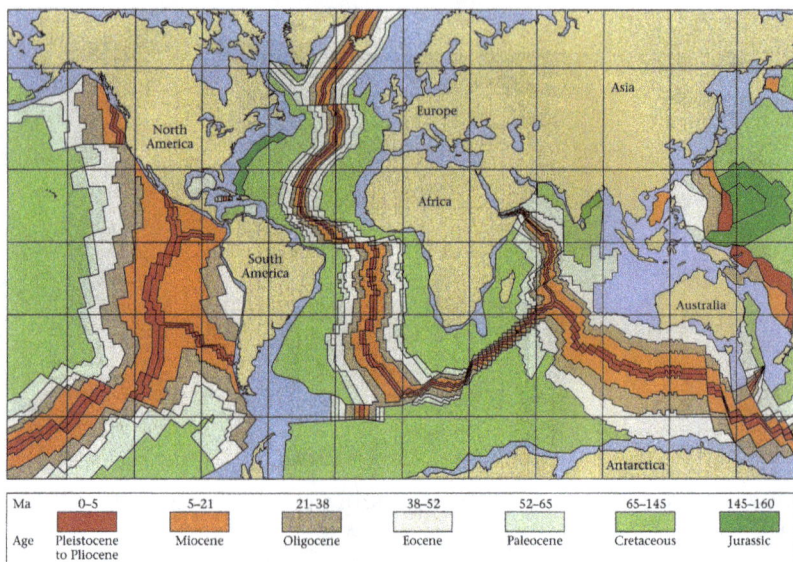

Distribution of the age of the floor of the world's oceans.

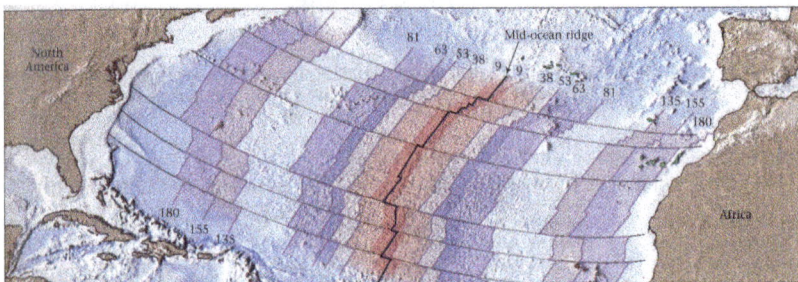

Detailed distribution of the age of the floor of the central Atlantic.

Figure 2.11. The present sea floor is young and has formed since the Jurassic. It becomes older with increasing distance from mid-ocean ridges (after Marshak, 2005).

ocean, which in turn provides a mechanism for the movement of continents or *continental drift*. These margins coincide with the mid-ocean ridges and the processes taking place here are shown in Fig. 2.16.

The position of Iceland is noteworthy; it straddles the Mid-Atlantic Ridge and this accounts for the frequency of volcanic eruptions on the island. It is the only place where processes occurring on the floor of the Atlantic can be observed and

Figure 2.12. The Mid-Atlantic Ridge. Note the position of Iceland on the ridge, which is cut by numerous offsets (National Geographic Magazine, 1968).

studied at the surface. Figure 2.17 depicts a steep-walled valley in the middle of the island, which represents the surface expression of the mid-Atlantic ridge.

Sagging and subsidence caused by normal faulting along the margins of the ocean create accommodation space, leading

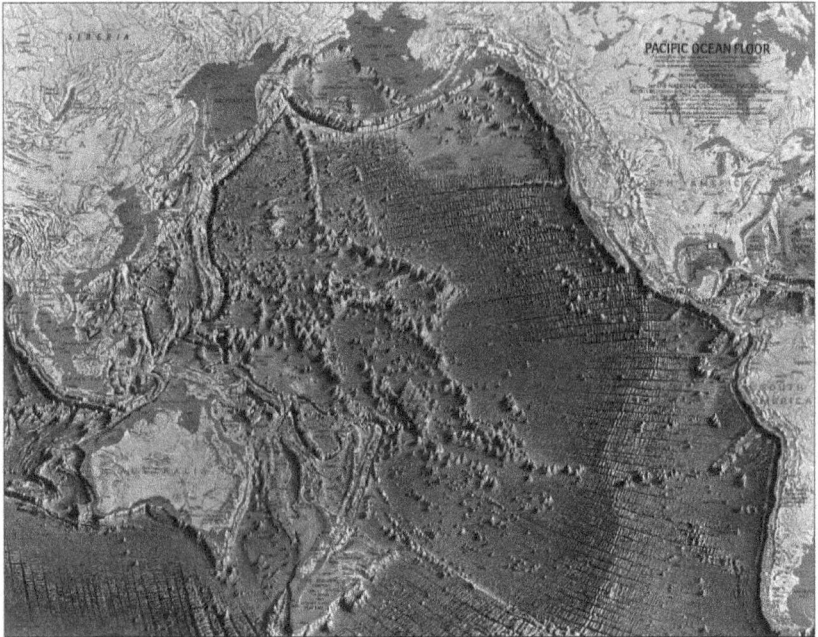

Figure 2.13. The Pacific Ocean ridge and trench systems (National Geographic Magazine, 1981).

to sediment accumulation (formation of sedimentary basins). The petroliferous basins along both the east and west coasts of the Atlantic were formed by this mechanism and are referred to as *passive margin* types. Being distant from zones of magmatic activity, passive margin basins are "cool"; i.e. they are characterised by relatively low geothermal gradients.

(b) *Destructive, consuming* or *convergent.* At these margins, also known as *subduction zones,* the crust formed at the accreting margins descends into the mantle. These margins coincide with deep-sea trenches and earthquake belts. In fact, the global distributions of these seismically active zones, presented in Fig. 2.18, were used by the pioneers of the plate tectonics theory to map the plate boundaries.

Figure 2.19 depicts the processes occurring at convergent margins. The underthrusting plate becomes heated as it descends

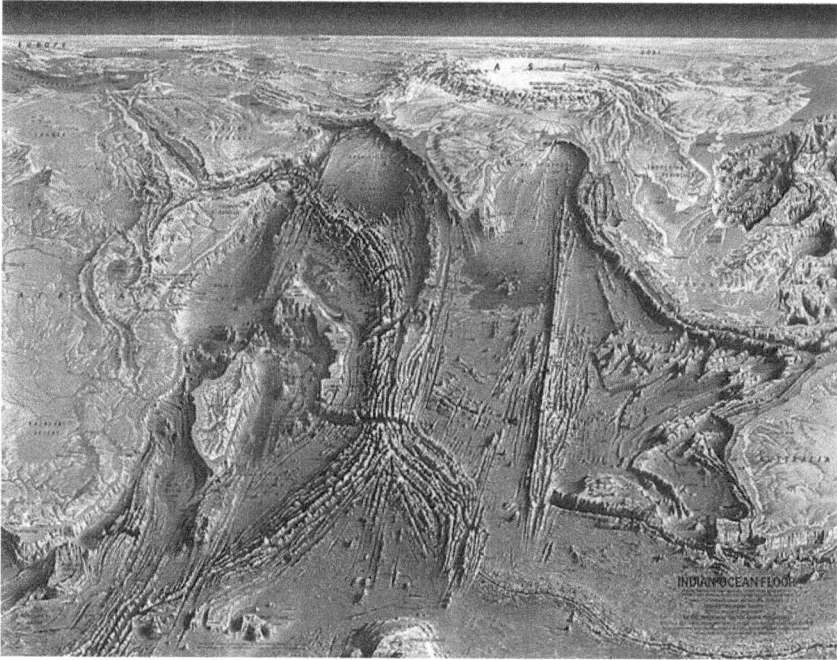

Figure 2.14. The Indian ocean ridge and trench systems (National Geographic Magazine, 1967).

into the mantle. It eventually melts, generating magma, which reaches the surface via volcanoes, causing metamorphism along the way. This explains the frequent association of volcanoes with subduction zones, as shown in Fig. 2.20. Volcanic activity also occurs along mid-ocean ridges as discussed above. However, these are zones of divergence and are unrelated to subduction and underthrusting.

The downward drag generated by the underthrusting plate leads to subsidence and sediment accumulation and thus sedimentary basin formation. Such basins are referred to as *active margin* types, examples of which occur in Southeast Asia. Being close to zones of volcanic activity, active margin basins are "hot"; i.e. they are characterised by relatively high geothermal gradients.

Figure 2.15. The Arctic Ocean ridge systems (National Geographic Magazine, 1971).

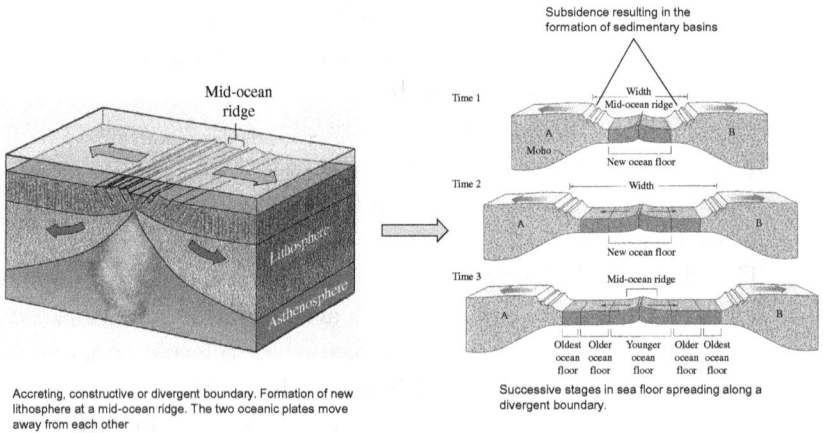

Figure 2.16. Processes taking place at a divergent boundary (after Marshak, 2005).

Figure 2.17. Steep-walled valley, representing the surface expression of the Mid-Atlantic Ridge in Iceland (from Sellwood, 2007).

(c) *Transform fault.* At these margins, the plates slip by each other and crust is neither created nor destroyed, as illustrated in Fig. 2.21. The best-known example of a transform fault margin on land is the San Andreas Fault in California, movements along which are the causes of the earthquakes experienced frequently in Los Angeles and San Francisco. It is an extension of the northeast–southwest-trending East Pacific Rise spreading ridge into the North American continent. The movement changes from spreading to strike-slip in the Gulf of California before continuing northwestward as shown in Fig. 2.22.

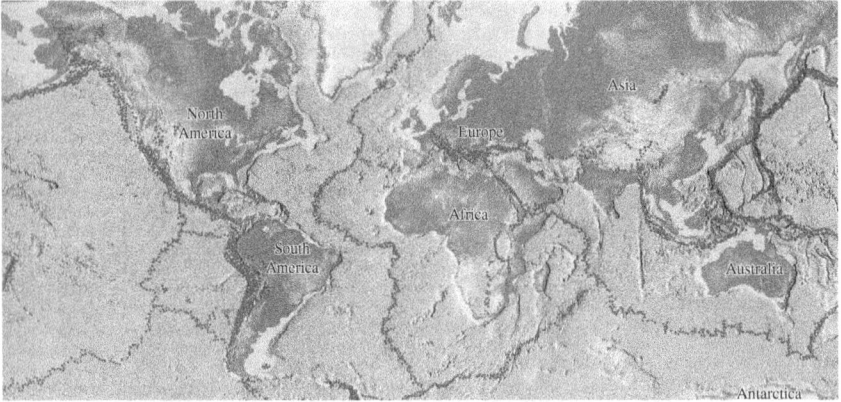

Figure 2.18. Earthquakes occur along distinct belts associated with plate margins. These belts have been used to map plate boundaries (after Marshak, 2005).

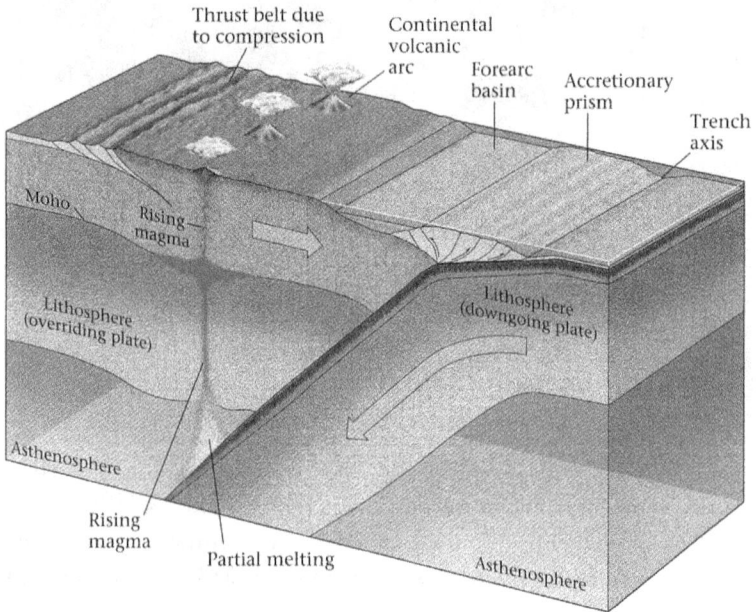

Figure 2.19. Destructive, consuming or convergent boundary. The crust created at mid-ocean ridges is destroyed as it descends into the mantle at subduction zones beneath deep-sea trenches. Melting of the descending plate generates magma, which reaches the surface via volcanoes (after Marshak, 2005).

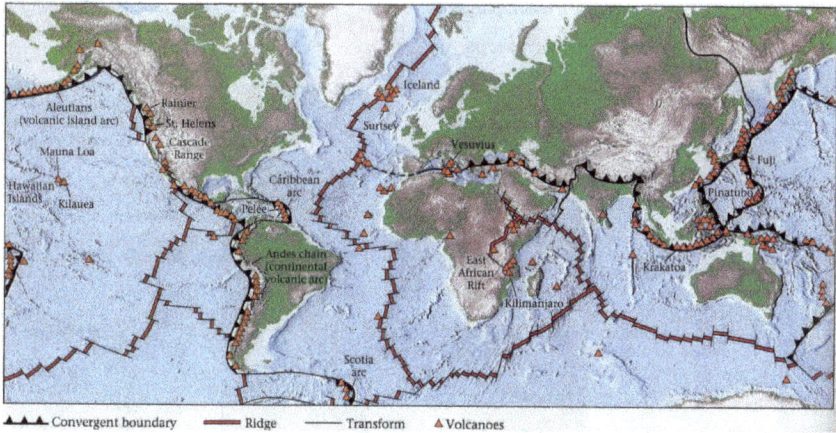

Figure 2.20. Global distribution of submarine and sub-aerial volcanoes. Volcanic activity occurs along zones of crustal divergence (mid-ocean ridges) and convergence (oceanic trenches) (after Marshak, 2005).

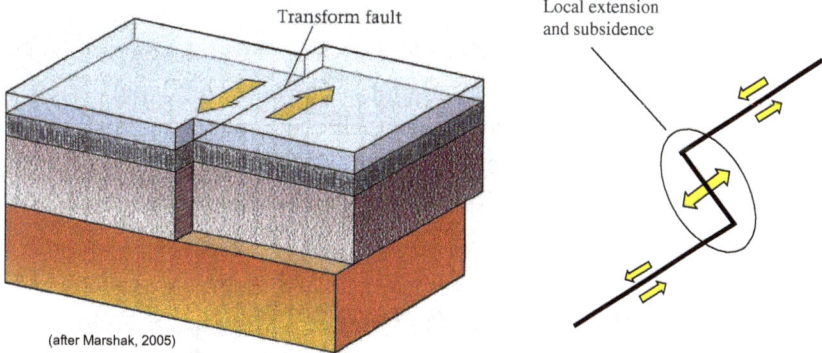

Figure 2.21. A transform fault plate boundary. An inflexion in the fault trend can cause local extension and subsidence.

Although the movement along transform faults is predominantly horizontal, inflections in the trend can cause local extension and subsidence, leading to the formation of sedimentary basins (Fig. 2.21). Being away from zones of volcanic activity, transform-fault-controlled basins are associated with relatively

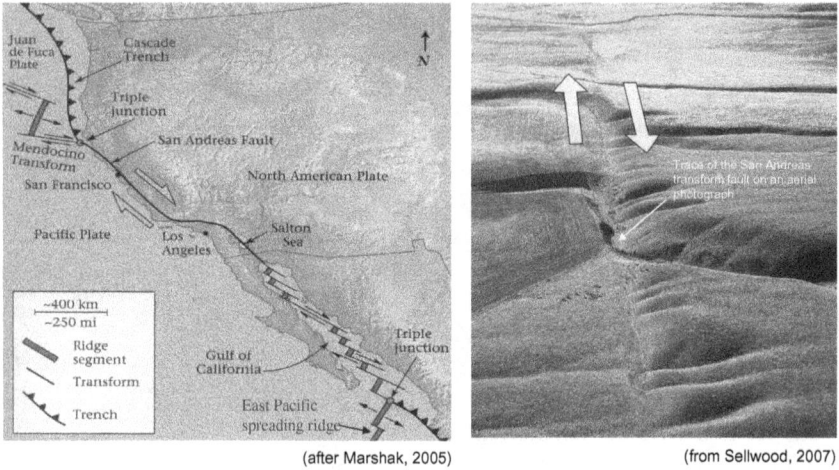

| (after Marshak, 2005) | (from Sellwood, 2007) |

Figure 2.22. The San Andreas Fault. An extension of the East Pacific spreading ridge into the North American continent. The movement changes from spreading to strike-slip in the Gulf of California, providing an example of a transform fault plate margin.

low geothermal gradients. The Los Angeles and Ventura basins of California fall into this category. They are richly petroliferous and it should be noted that California was the leading oil-producing state in the US in the 1920s until the discovery of the major oil fields of Texas.

## 2.4.2. *Influence of plate tectonic setting on sedimentary basin characteristics*

The differences in plate tectonic setting result in differences in basin characteristics such as tectonic regime (tensional in the case of basins associated with crustal divergence and compressional in the case of basins related to crustal convergence), heat flow (low in basins in divergent and transform fault settings and high in those in convergent settings), lithology of the basin fill, source rock maturity depth and the nature of the hydrocarbons (lower temperatures favour the formation of heavier oils while higher temperatures lead to the dominance of lighter oils and gas).

### 2.4.3.  *Distribution of fossil faunae, florae and other geological features*

Today, different faunae and florae are found on different continents, but there are striking similarities among the fossil faunae, florae and certain geological features between landmasses that are now separated by large oceans. None of these species could have traversed the ancient oceans to account for the similarities, which leads to the conclusion that the ancient continents must have constituted a single large landmass or supercontinent in the past. The breakup of this supercontinent and the drifting apart of its fragments must have therefore occurred. The terrestrial fossil species can be matched by restoring the continental fragments to their original pre-drift position as illustrated in Fig. 2.23. The same approach explains the continuity of some geological features between landmasses that are now widely separated as shown in Fig. 2.24. Moving continents or continental

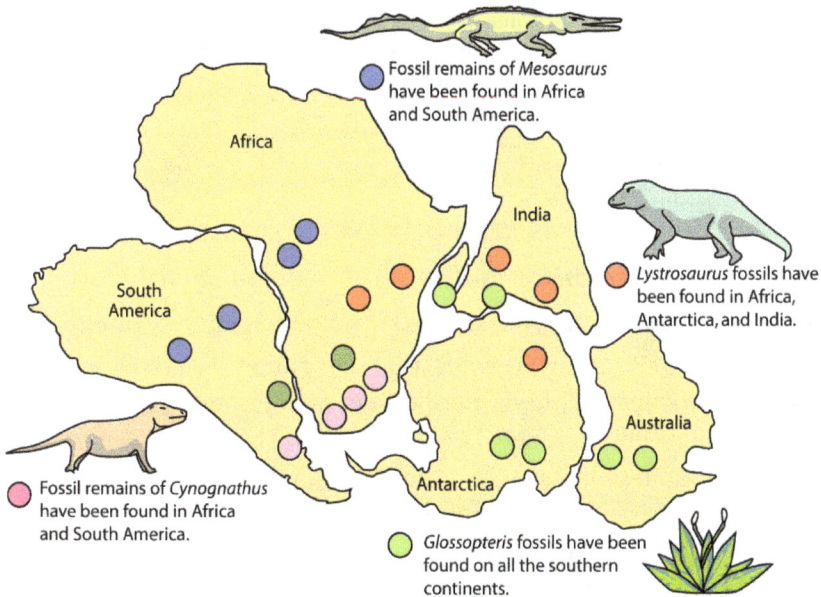

Figure 2.23. Matching fossil faunae and florae by restoring the continents to their pre-drift positions (redrawn from Marshak, 2005).

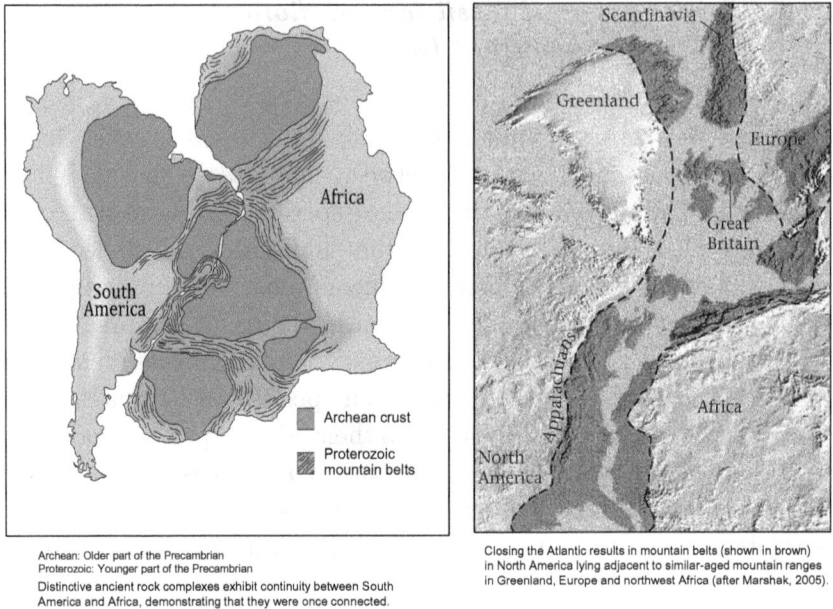

Archean: Older part of the Precambrian
Proterozoic: Younger part of the Precambrian
Distinctive ancient rock complexes exhibit continuity between South
America and Africa, demonstrating that they were once connected.

Closing the Atlantic results in mountain belts (shown in brown)
in North America lying adjacent to similar-aged mountain ranges
in Greenland, Europe and northwest Africa (after Marshak, 2005).

Figure 2.24. Continuity of some geological features between continents on the
opposite sides of the Atlantic once the landmasses have been restored to their
pre-drift positions (redrawn from Marshak, 2005).

drift and sea-floor spreading, its corollary, can thus be placed in the
context of plate tectonics.

### 2.4.4. *Plate tectonics and mountain formation*

When the floor of an ocean separating continents is consumed at a
subduction zone, the continents that are riding on plates converge
and collide. Due to its lower density and greater buoyancy, continental crust does not subduct, i.e. descend into the underlying mantle.
The rocks and sediments that lie between the continents become
squeezed and thrown up as a crumpled pile, forming mountain belts.
This is the mechanism by which mountain chains have been formed
throughout the geological history of the Earth.

The formation of the great Alpine-Himalayan ranges serves as
a recent example of mountain building. In the Late Triassic, about
200 million years ago, Africa and India were part of a supercontinent

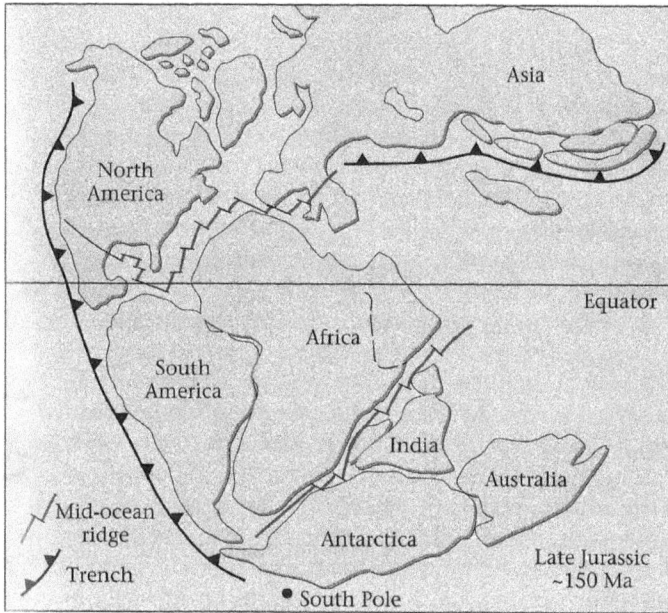

Figure 2.25. The Gondwana landmass (South America, Africa, India, Australia and Antarctica) in the late Jurassic (after Marshak, 2005).

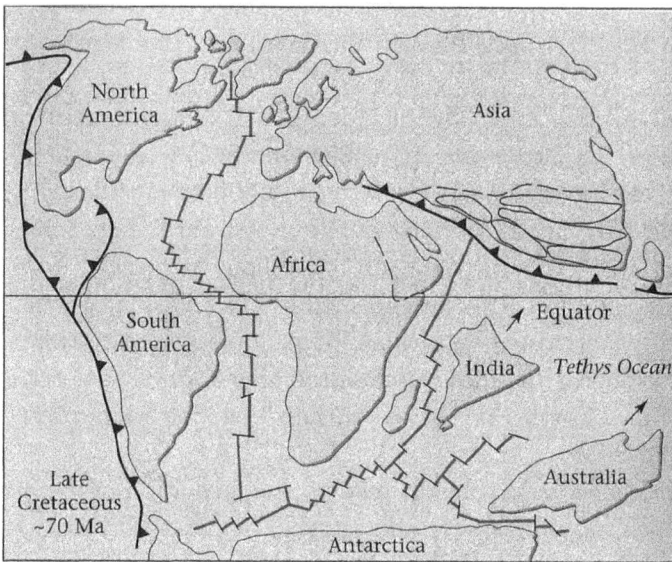

Figure 2.26. Restoration of the continents to their Late Cretaceous positions (after Marshak, 2005).

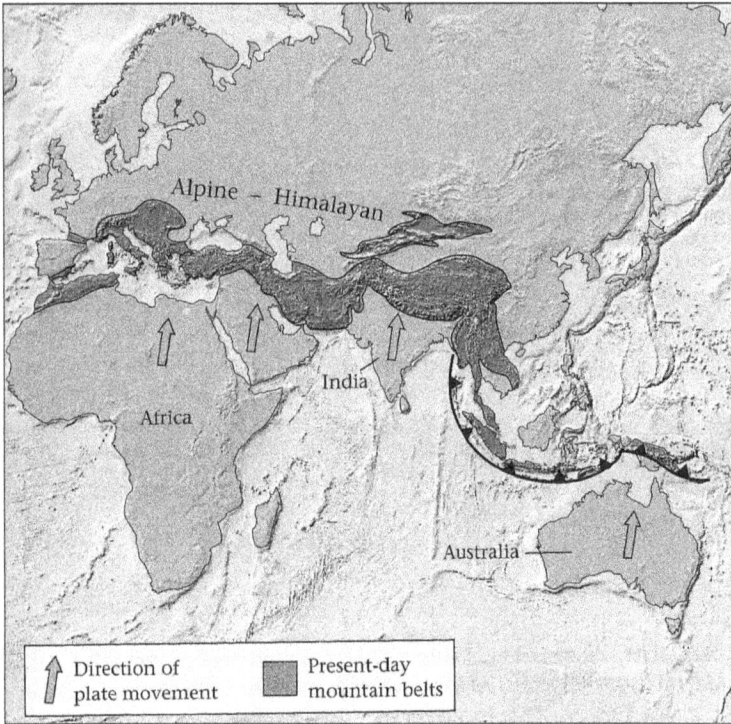

Figure 2.27. The Alpine-Himalayan mountain belt (after Marshak, 2005).

called Gondwana that also included South America, Australia and Antarctica, as shown in Fig. 2.25. Gondwana fragmented in the Jurassic and by the Late Cretaceous, about 70 million years ago, India and Africa were drifting rapidly northward (Fig. 2.26).

India collided with Asia 40–50 million years ago, forming the Himalayas, and the Alps were formed by the collision between Africa and Europe at about the same time. This was a major tectonic event, forming a mountain belt stretching from Southeast Asia to Europe and North Africa, as outlined on the map presented in Fig. 2.27.

Convergence and collision lead to the formation of a supercontinent, which subsequently breaks up; the fragments drift apart only to recombine later and form a new supercontinent. The junction along which the continents become welded together after collision is called

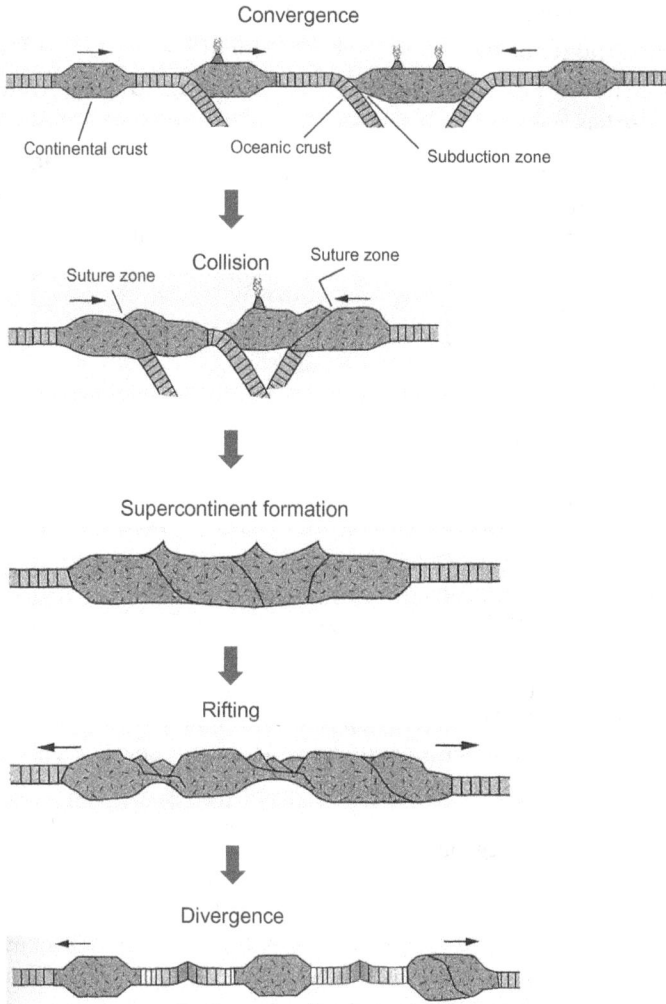

Figure 2.28. Diagrammatic illustration of the supercontinent cycle (after Marshak, 2005).

a *suture zone*. Supercontinent creation and fragmentation is a cyclic event that has occurred many times in the geologic past and will recur in the future. Figure 2.28 is a diagrammatic representation of the process; the duration of each cycle is estimated to be about 500 million years. Current projections indicate that today's shifting continents are heading for collision 250 million years from now, which will create the next supercontinent. This will be similar to Pangaea, which fragmented in the Early Triassic, leading to the formation of the present continents.

Earth scientists now look back 750 million years into the past and model the shape of the Earth's far distant future using a computer. Ultimately, however, all supercontinents exist only in the human imagination, but understanding the 'supercontinent cycle' throws light on how our planet works.

## References

Allen, P.A. and Allen, J.R. (2005). *Basin Analysis: Principles and Applications, 2$^{nd}$ Edition*, Blackwell Publishing, London, UK.

Arctic Ocean Floor Map (1971). National Geographic Magazine. Available at: http://www.natgeomaps.com/arctic-ocean-floor-map.

Atlantic Ocean Floor Map (1968). National Geographic Magazine. Available at: http://www.natgeomaps.com/atlantic-ocean-floor-map.

Beydoun, Z.R. (1991). Arabian plate hydrocarbon geology — a plate tectonic approach, *AAPG Studies in Geology, No. 33*.

Catuneanu, O. (2006). *Principles of Sequence Stratigraphy, 1$^{st}$ Edition*, Elsevier Scientific Publishing Company, London, UK.

Chapman, R.E. (1977). "Petroleum exploration and development", in *Our Industry Petroleum*, British Petroleum Company Ltd, London, UK.

Indian Ocean Floor Map (1967). National Geographic Magazine. Available at: http://www.natgeomaps.com/indian-ocean-floor-map.

Kirby, B.E. (1977). "Oil producing countries of the world", in *Our Industry Petroleum*, British Petroleum Company Ltd, London, UK.

Konert, G., Afifi, A.M., Al-Hajri, S.A. *et al.* (2001). Paleozoic stratigraphy and hydrocarbon habitat of the Arabian plate, *Pratt II Conference*, AAPG, Tulsa, OK.

Marshak, S. (2005). *Earth: Portrait of a Planet, 2$^{nd}$ Edition*, W.W. Norton, New York, NY.

Pacific Ocean Floor Map (1983). National Geographic Magazine. Available at: http://www.natgeomaps.com/pacific-ocean-floor-map.

Sellwood, B.W. (2007). *Plate Tectonics, Basins and Heat*, Department of Earth Science and Engineering, Imperial College, London.

Ziegler, P.A. (1990). *Geological Atlas of Western and Central Europe, 2$^{nd}$ Edition*, Shell International Petroleum Maatschppij B.V., The Netherlands.

# Chapter 3

# Chemical Composition of Petroleum

## 3.1. Introduction

The term *petroleum* encompasses both crude oil and gas, which are composed almost entirely of *hydrocarbons*, compounds formed from hydrogen and carbon. The compounds present in petroleum include:

- alkanes,
- alkenes,
- cycloalkanes or cycloparaffins,
- aromatics,
- sulphur compounds,
- nitrogen, oxygen and metallic compounds.

In all these compounds carbon has a valency of 4 and hydrogen 1. The relative proportions of these compounds determine the physical properties (density, viscosity, pour point, etc.) of petroleum.

In continental Europe, crude oil density is expressed in terms of specific gravity (SG), while in the US and UK a scale devised by the American Petroleum Institute, the API gravity, is used. API gravity and SG are linked by the following empirical relationship:

$$\text{API gravity} = (141.5/\text{SG at } 60°\text{F}) - 131.5. \qquad (3.1)$$

SGs of crude oils range from 0.8 gm/cc to 1.0 gm/cc, corresponding to 45° API to 10° API. The higher the API gravity, the lower the

Table 3.1. API gravity ranges of the various types of crude oils.

| Category | API gravity |
|---|---|
| Light oils | Above 31.1° |
| Medium oils | 22.3–31.1° |
| Heavy oils | 10–22.3° |
| Extra heavy oils | Below 10° |

density and viscosity and hence the better the quality of the oil. Broadly, oils are categorised as light, medium, heavy and extra heavy. The API gravity ranges corresponding to these categories are listed in Table 3.1.

Petroleum consists of complex mixtures of hydrocarbons and non-hydrocarbons and the substances present in crude oil and gas are described briefly below.

## 3.2.  Alkanes

*Alkanes* are saturated paraffins, characterised by a straight-chain molecular structure and single carbon-to-carbon bonds. They have a general formula $C_nH_{2n+2}$ and the simplest paraffin is methane, $CH_4$. The lower-molecular-weight paraffins are listed in Table 3.2. Alkanes up to $C_4$, i.e. those with up to four carbon atoms, are gaseous at room temperature, liquid between $C_5$ and $C_{15}$ and solid (wax) in the case of those with more than 15 carbon atoms. Density, viscosity and boiling point increase with increasing number of carbon atoms.

The straight-chain alkanes are known as *normal* paraffins, abbreviated to *n*-paraffins. In some cases the alkane chains are branched, as shown in Fig. 3.1. These compounds are referred to as *isomers* and the corresponding paraffins are called *iso*-paraffins. Isomers have the same molecular formula, but have a different arrangement of the atoms in space and are characterised by different chemical and physical properties. For example, the boiling point of an *n*-paraffin is higher than that of an *iso*-paraffin with the same molecular formula.

Table 3.2. The lower-molecular-weight paraffins.

| Name | Formula | Structure | State (at room temp.) |
|------|---------|-----------|------------------------|
| Methane | $CH_4$ | $\begin{array}{c} H \\ | \\ H-C-H \\ | \\ H \end{array}$ | Gas |
| Ethane | $C_2H_6$ | $\begin{array}{cc} H & H \\ | & | \\ H-C-C-H \\ | & | \\ H & H \end{array}$ | Gas |
| Propane | $C_3H_8$ | $\begin{array}{ccc} H & H & H \\ | & | & | \\ H-C-C-C-H \\ | & | & | \\ H & H & H \end{array}$ | Gas |
| Butane | $C_4H_{10}$ | $\begin{array}{cccc} H & H & H & H \\ | & | & | & | \\ H-C-C-C-C-H \\ | & | & | & | \\ H & H & H & H \end{array}$ | Gas |
| Pentane | $C_5H_{12}$ | $\begin{array}{ccccc} H & H & H & H & H \\ | & | & | & | & | \\ H-C-C-C-C-C-H \\ | & | & | & | & | \\ H & H & H & H & H \end{array}$ | Liquid |
| Hexane | $C_6H_{14}$ | $\begin{array}{cccccc} H & H & H & H & H & H \\ | & | & | & | & | & | \\ H-C-C-C-C-C-C-H \\ | & | & | & | & | & | \\ H & H & H & H & H & H \end{array}$ | Liquid |

Paraffins are the most abundant hydrocarbons in both gaseous and liquid petroleums.

## 3.3. Alkenes

Also known as *olefins*, *alkenes* are unsaturated straight-chain hydrocarbons, containing at least one carbon-to-carbon double bond as shown in Fig. 3.2. They have a general formula $C_nH_{2n}$ and the physical properties of alkenes are comparable with those of alkanes. At room temperature, the lower molecular weight alkenes — ethene, propene and butene — are gases; those containing 5–16 carbon atoms are liquids, and higher members of the series are waxy solids. Alkenes also show isomerism and Fig. 3.3 shows the comparison between *n*-butene and *iso*-butene.

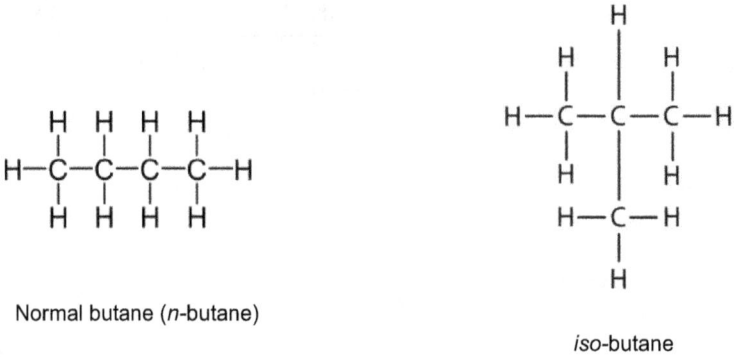

Normal butane (*n*-butane)

iso-butane

Figure 3.1. Comparison of the molecular structures of normal butane, $C_4H_{10}$, and pentane, $C_5H_{12}$, and their isomers.

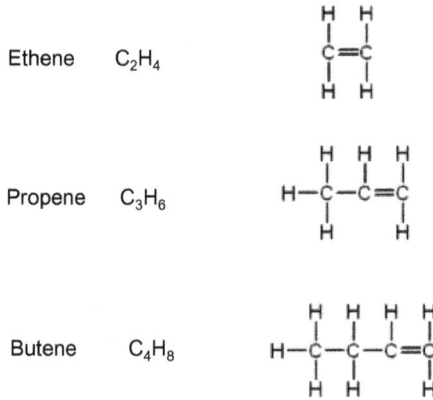

Ethene    $C_2H_4$

Propene    $C_3H_6$

Butene    $C_4H_8$

Figure 3.2. Lower-molecular-weight alkenes.

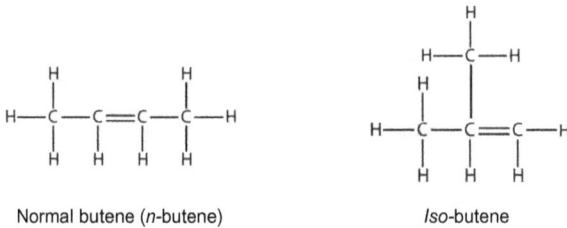

Normal butene (*n*-butene)

Iso-butene

Figure 3.3. Comparison of the molecular structures of normal butene, $C_4H_8$ and *iso*-butene.

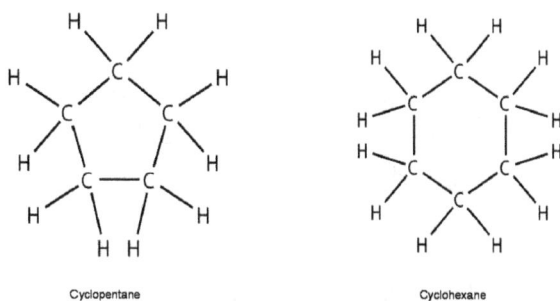

Cyclopentane

Cyclohexane

Figure 3.4. Molecular structures of cyclopentane, $C_5H_{10}$, and cyclohexane, $C_6H_{12}$.

Alkenes represent only a minor constituent of petroleum.

## 3.4. Cycloalkanes

*Cycloalkanes* are also referred to as *cycloparaffins* or *naphthenes* and are saturated closed-ring hydrocarbons with the general formula $C_nH_{2n}$. The simplest cycloparaffin is cyclopropane, $C_3H_6$, but in crude oils the most common cycloparaffins are cyclopentane, $C_5H_{10}$, and cyclohexane, $C_6H_{12}$, the molecular structures of which are shown in Fig. 3.4.

Cycloparaffins with up to four carbon atoms are gaseous at room temperature but all the other monocyclic naphthenes are liquid. The naphthenes constitute an important portion of petroleums as well as of most products.

## 3.5. Aromatics

*Aromatics* are unsaturated closed-ring hydrocarbons with the general formula $C_nH_{2n-6}$. The benzene ring, $C_6H_6$, is the building block of the aromatic component of crude oil. It is characterised by alternating single and double bonds between the six carbon atoms and each carbon atom is linked to a hydrogen atom, as illustrated in Fig. 3.5. Benzene is a colourless and volatile liquid at room temperature and has many derivatives. Aromatics rarely amount to more than 15% of crude oils.

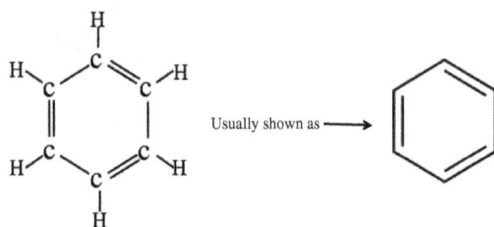

Figure 3.5. The benzene ring, $C_6H_6$.

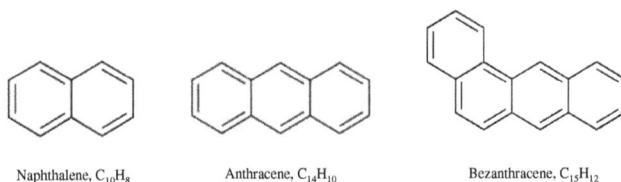

Naphthalene, $C_{10}H_8$        Anthracene, $C_{14}H_{10}$        Bezanthracene, $C_{15}H_{12}$

Figure 3.6. Aromatic compounds formed by sharing two or more benzene rings.

A great variety of aromatic compounds exist in petroleum. They are formed by sharing two or more benzene rings as illustrated in Fig. 3.6.

## 3.6.  Sulphur Compounds

Sulphur is a common constituent of most crude oils and gases. Typically, the content is in the range of 0.1–5.5% by weight and is in the form of free sulphur, hydrogen sulphide, $H_2S$ or organic sulphur compounds. The best-known sulphur compound in petroleum is $H_2S$, which is characterised by a pungent odour. All sulphur compounds are corrosive and poisonous, making the handling of the petroleum more difficult. Their presence lowers the quality and hence the price of the oil and gas. Hydrocarbons with low sulphur content are referred to as "sweet" oil or gas, while those with a high proportion of sulphur are called "sour" oil or gas. In general, an increase in sulphur content is accompanied by an increase in SG and a decrease in API gravity, as shown in Fig. 3.7.

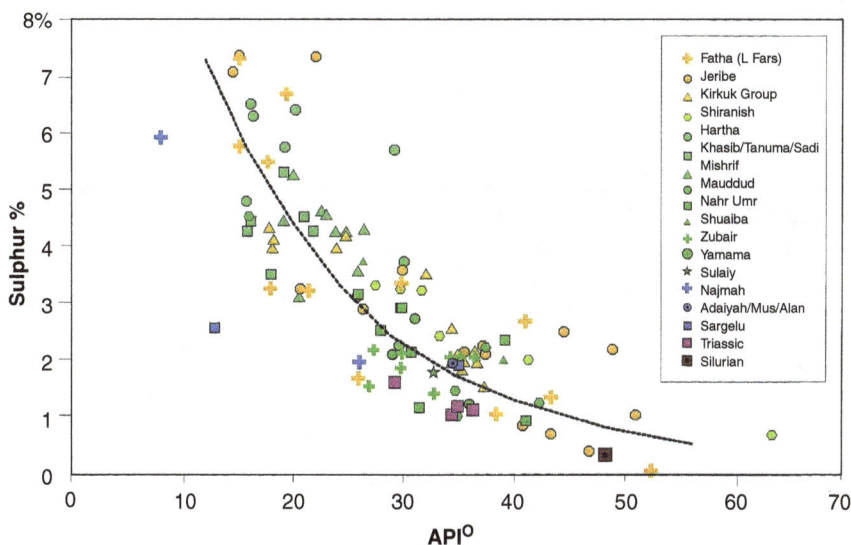

Figure 3.7. Relationship between sulphur content and API gravity of oils from various fields in Iraq (after Jassim and Al-Gailani, 2006; courtesy Dr M.B. Al-Gailani, 2006).

Pennsylvanian, North Sea, North African and Nigerian crudes are examples of light, sweet oils while Mexican crudes fall in the heavy, sour category, with sulphur contents of 3–5% by weight.

## 3.7. Nitrogen, Oxygen and Metallic Compounds

Nitrogen, oxygen and metallic compounds are present only in minor quantities in petroleum. The nitrogen content of crude oil ranges between 0.01% and 0.9% by weight and most of the compounds containing nitrogen are characterised by high boiling points. They are the products of the breakdown of chlorophyll molecules and often contain a metal, commonly nickel or vanadium.

The proportion of oxygen in petroleum varies from 0.06% to 0.4% by weight. It occurs in crude oil in the form of organic acids, which range from simple straight chain molecules to complex compounds containing between one and thirty carbon atoms. The former include alkane acids while the latter consists of naphthenic and naphthenic-aromatic acids. Crude oil boiling point rises with increasing proportions of compounds containing oxygen.

## Reference

Jassim, S.Z. and Al-Gailni, M.B. (2006). "Hydrocarbons", in Jassim, S.Z. and Goff, J.C. (eds.), *Geology of Iraq*, Dolin, Prague and Moravian Museum, Brno, Czech Republic.

# Chapter 4

# Petroleum System Analysis

## 4.1. Introduction

Oil and gas fields are economically viable concentrations of hydro-carbons, the formation of which is controlled by a number of factors referred to as *elements* and *processes*. *Petroleum system analysis* is the study of how these elements and processes interact to create hydrocarbon-bearing provinces. It should be emphasised that this discussion applies to "conventional" accumulations, i.e. those where the hydrocarbons are stored in voids in certain types of rocks in oil and gas fields. The elements and processes are reviewed briefly below and covered in more detail in the ensuing sections.

### 4.1.1. *Elements*

- A *source rock* is a rock containing sufficient hydrocarbon-prone organic matter to generate oil and gas.
- A *reservoir rock* is a rock in which oil and gas accumulate. *Porosity* and *permeability* are fundamental properties of reservoir rocks. Porosity is the space between rock grains in which oil and gas accumulate and permeability is defined as the ease with which oil and gas can move through the pore spaces between the grains.
- A *seal* is a rock through which oil and gas cannot move effectively over time (such as mudstone or evaporite).
- The *migration route* is the avenue through which oil and gas move from the source into the reservoir rock to the trap.

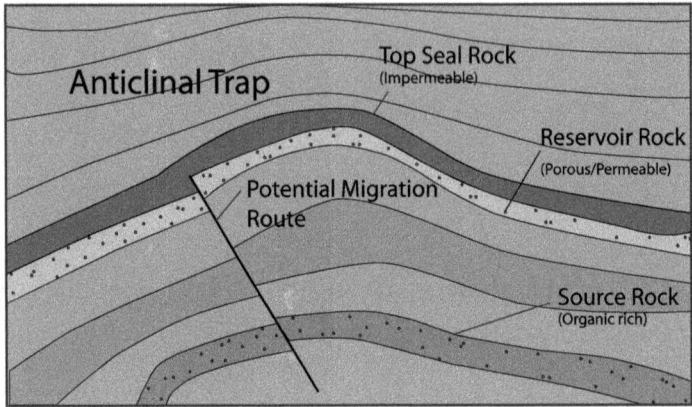

Figure 4.1. Diagrammatic illustration of the elements of a petroleum system.

- The *trap* is a feature that arrests the migration process and causes oil and gas to accumulate.

Figure 4.1 provides a diagrammatic illustration of the elements.

### 4.1.2.  *Processes*

- *Generation* results from the burial of the source rock to sufficient depths, which subjects it to high temperatures that convert the organic matter into hydrocarbons.
- *Migration* is the movement of the hydrocarbons out of the source rock into a reservoir to form accumulations.
- *Accumulation* is the concentration of hydrocarbons in a trap to form oil and gas fields.
- *Timing*. Trap formation should predate or at least be simultaneous with hydrocarbon migration. Traps formed after migration are invariably dry.

Figure 4.2 presents a diagrammatic illustration of the processes.

### 4.1.3.  *Significance of oil and gas seeps*

*Seeps* are the results of leaks from an existing accumulation. They do not originate directly from source rocks since the generation

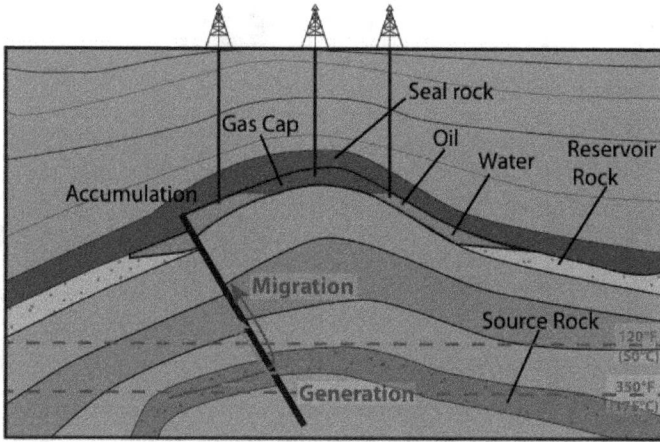

Figure 4.2. Diagrammatic illustration of the processes in a petroleum system.

and migration processes are too slow to allow the escape of the hydrocarbons to the surface without undergoing accumulation first.

Existence of an accumulation means that the conditions for the occurrence of hydrocarbons have been satisfied:

- All the elements (source, reservoir trap and seal) are present in the area.
- All the processes (generation, migration and accumulation) have taken place.

Seeps fall into two categories, namely, *active* and *inactive*. In the case of active seeps, oil and gas flows can be observed directly at the surface. Figure 4.3 shows crude oil bubbling to the surface in a ditch in southwest Iran, and Fig. 4.4 presents a burning gas seep in northern Iraq, an example of the "eternal fires" mentioned in ancient historical chronicles. An offshore gas seep, in the North Sea, is depicted in Fig. 4.5. Active oil and gas seeps are generally considered to be a positive factor; their presence enhances the hydrocarbon potential of the area.

*Inactive* or *dead* seeps occur in the form of heavy oil or bitumen impregnations, an example of which from northern Greenland is

Figure 4.3.  An active oil seep in southwest Iran (photo by M. Ala).

Figure 4.4.  An active gas seep, Kirkuk area, northern Iraq (after Aqrawi *et al.*, 2010).

Figure 4.5. An escape of natural gas in the North Sea (after Cuddington and Lowther, 1977).

shown in Fig. 4.6. They indicate that oil was once present in the area, but whether any of it has been preserved remains uncertain.

## 4.2.   Source Rocks and Generation of Petroleum

In conventional oil and gas fields the hydrocarbons are not indigenous to the reservoir. They are generated from the organic matter in the source rock, and migrate into the reservoir. Source rocks are very fine-grained, dark-coloured sediments such as shale or carbonate mudstone. The dark colour is due to the presence of the organic matter. Figure 4.7 shows an exposure of an organic rich shale in along the Yorkshire coast, northeast England.

The preservation of the organic matter in the source rock is therefore a key factor and this leads to a discussion of the conditions necessary for the deposition of organic-rich sediments. The requirements are:

Figure 4.6. Dead or inactive seep. Bitumen-impregnated sandstone, northern Greenland (after Henriksen, 2008).

Figure 4.7. Exposure of an organic-rich shale: Lower Jurassic, Yorkshire coast, NE England (photo by M. Ala).

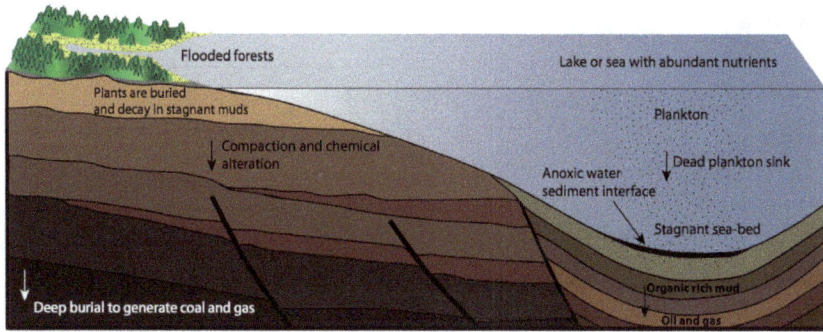

Figure 4.8. Illustration of ideal conditions for the deposition and preservation of organic matter (modified and redrawn from UKOOA).

- a subaqueous environment;
- oxic surface waters to support abundant organic life, from which the hydrocarbons are ultimately derived; and
- stagnant sea-bed conditions, leading to the development of an anoxic water–sediment interface. This is essential, otherwise the organic matter becomes oxidised into $CO_2$ and water and will not be preserved.

Figure 4.8 illustrates the ideal conditions for the preservation of the organic matter and Fig. 4.9 depicts the later stages of the generation, migration and accumulation processes.

For stagnant sea-bed conditions to develop, an enclosed environment with restricted circulation is the primary requirement. Identifying and studying modern analogues of such environments are useful in the understanding of sites of ancient source-rock deposition. Such modern analogues include the Mediterranean and Black Seas; both are silled, enclosed basins with restricted access to open waters. The Strait of Gibraltar separates the Mediterranean from the open waters of the Atlantic while the Strait of Bosphorus restricts the connection between the Black Sea and the Mediterranean. However, as shown in Fig. 4.10, the Strait of Gibraltar is not an effective barrier to circulation. Also, evaporation at the eastern end of the basin increases the concentration and density of the water, which sinks and aids the circulation process. Consequently, the Mediterranean is an

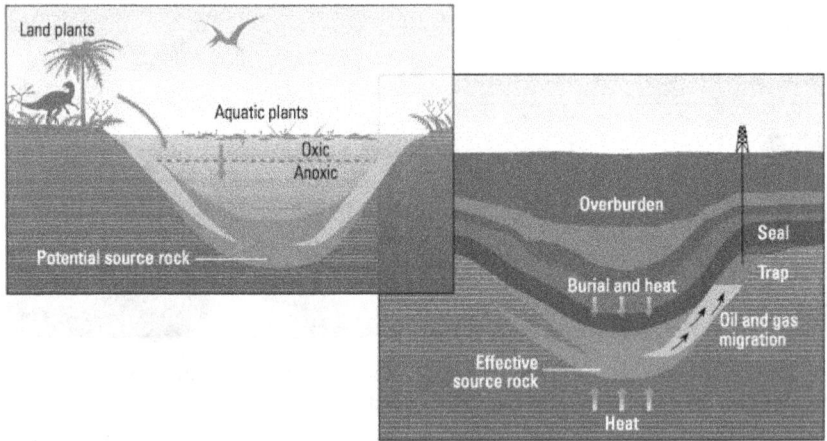

Figure 4.9. Illustration of geological and thermal effects on a source rock resulting in the generation, migration and entrapment of hydrocarbons (copyright Schlumberger, after Al-Hajeri *et al.*, 2009, used with permission).

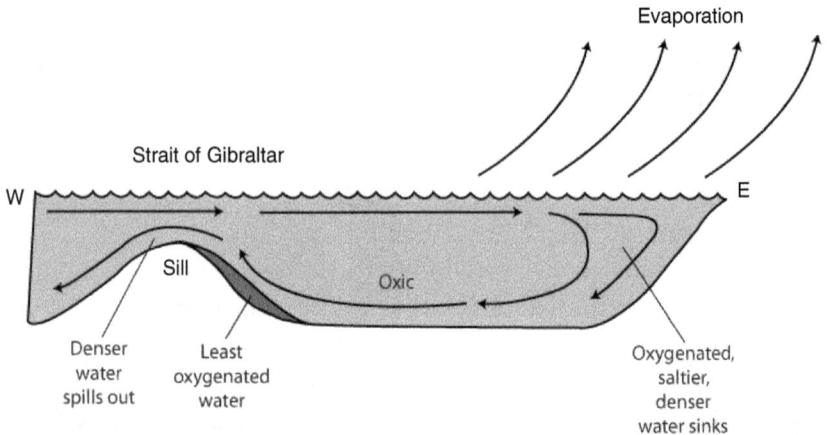

Figure 4.10. The Mediterranean Sea is a present-day oxic basin with poor organic matter preservation potential (modified and redrawn from Hunt, 1995).

oxic basin, which means that it has poor organic matter preservation potential. It is not, therefore, a suitable analogue for ancient sites of source-rock deposition.

Being only 27 m deep, the Bosphorus sill is an effective barrier to circulation in the Black Sea, creating an anoxic environment, as

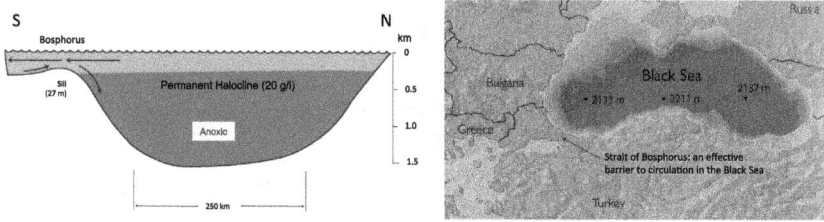

Figure 4.11. The Black Sea is a present-day anoxic basin with good organic matter preservation potential.

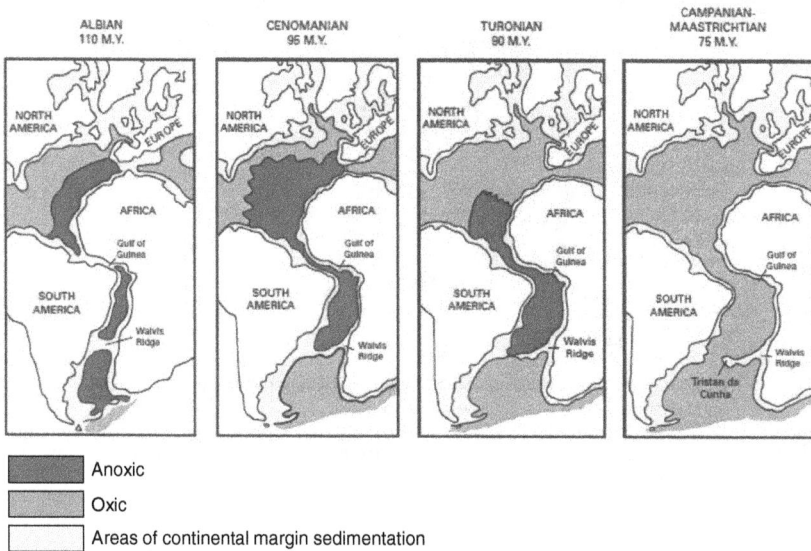

Figure 4.12. Anoxic basins developing in restricted seaways after the Early Cretaceous breakup of Africa and south America. These areas became the site of deposition of the source rocks of the petroliferous basins along the Atlantic coasts of Africa and South America (after Brownfield & Charpentier, 2006).

illustrated in Fig. 4.11. Organic-rich sediments are accumulating here now and the anoxic conditions favour the preservation of the organic matter, making the Black Sea a suitable analogue for ancient sites where source-type rocks were deposited.

Figure 4.12 shows the development of anoxic basins in the central Atlantic following the Late Cretaceous breakup of Africa and South America.

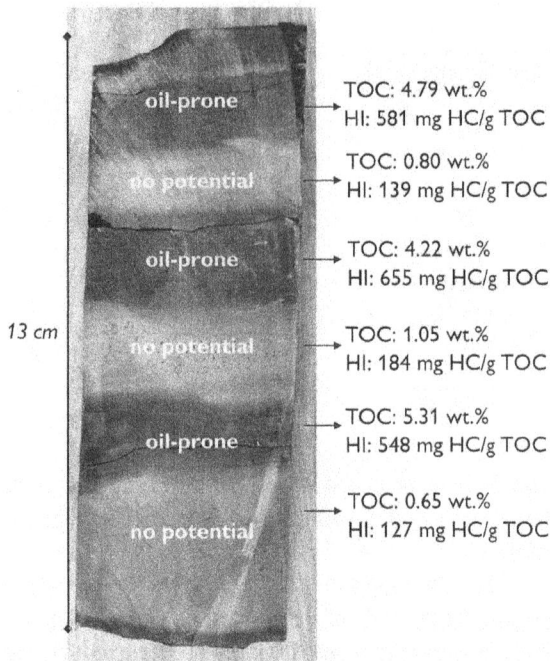

TOC: 4.79 wt.%
HI: 581 mg HC/g TOC

TOC: 0.80 wt.%
HI: 139 mg HC/g TOC

TOC: 4.22 wt.%
HI: 655 mg HC/g TOC

TOC: 1.05 wt.%
HI: 184 mg HC/g TOC

TOC: 5.31 wt.%
HI: 548 mg HC/g TOC

TOC: 0.65 wt.%
HI: 127 mg HC/g TOC

TOC: total organic content

HI: hydrogen index, indicative of the amount of hydrogen in the organic matter.
High HI indicates greater oil generation potential.

Figure 4.13. Core section, showing the occurrence of organic matter in bands in a mudstone, offshore Vietnam (after Petersen *et al.*, 2014).

The organic matter often occurs in bands in sediments, as shown in Fig. 4.13. The darker the colour the higher the total organic content, abbreviated to TOC.

It is estimated that 10–20% of petroleum is formed directly from the hydrocarbons produced by living organisms. The remaining 80–90% of oil and gas is formed by the thermal alteration of *kerogen*, the organic matter in sedimentary rocks that is insoluble in organic solvents. The most important factor in the generation of hydrocarbons from kerogen is temperature. For temperature to rise to the required level, the source rock must be buried. As the temperature increases, the kerogen *matures*. With increasing maturity, the colour of the kerogen changes from yellow to brown,

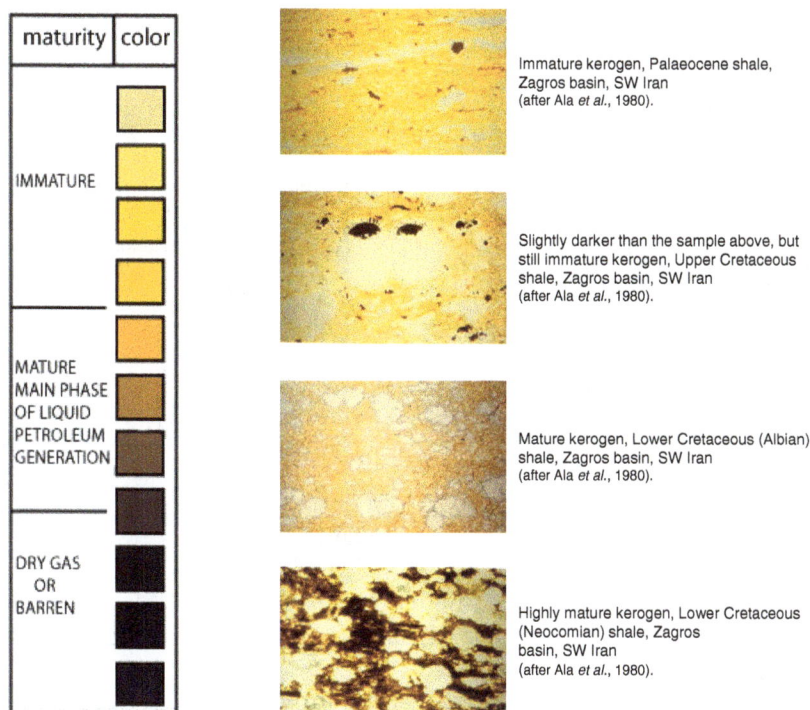

| maturity | color |
|----------|-------|

IMMATURE

MATURE
MAIN PHASE
OF LIQUID
PETROLEUM
GENERATION

DRY GAS
OR
BARREN

Immature kerogen, Palaeocene shale,
Zagros basin, SW Iran
(after Ala *et al.*, 1980).

Slightly darker than the sample above, but
still immature kerogen, Upper Cretaceous
shale, Zagros basin, SW Iran
(after Ala *et al.*, 1980).

Mature kerogen, Lower Cretaceous (Albian)
shale, Zagros basin, SW Iran
(after Ala *et al.*, 1980).

Highly mature kerogen, Lower Cretaceous
(Neocomian) shale, Zagros
basin, SW Iran
(after Ala *et al.*, 1980).

Figure 4.14. Changes in the colour of kerogen with increasing maturity.

to dark brown and finally black. The changes in colour and the corresponding kerogen maturity levels are illustrated in Fig. 4.14.

Most authorities agree that the threshold temperature for the onset of maturity is 75−80°C. Initially, heavy oil is generated followed by light oil, then "wet" gas (oil and gas together) and finally dry gas, as depicted in Figs. 4.15 and 4.16. The oil- and gas-generation phases of the process are referred to respectively as the oil and gas windows, with an overlapping segment representing wet gas formation. Hydrocarbon generation ceases once the hydrogen content of the kerogen is exhausted.

### 4.2.1. *Types of kerogen*

Kerogen is commonly categorised into four types and these are described briefly below.

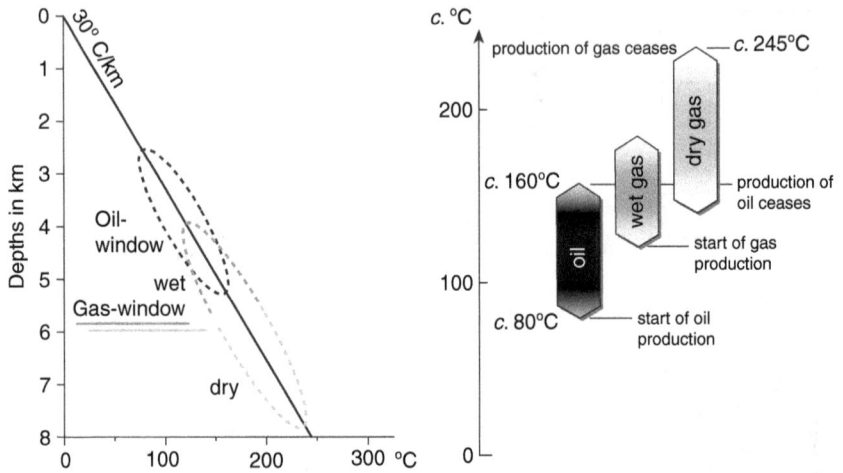

Figure 4.15. Influence of temperature and depth on hydrocarbon generation, assuming an average geothermal gradient of $30°C/km$ (after Henriksen, 2008).

- *Type I* is derived mainly from algae. It is commonly lacustrine in origin (deposited in fresh water lakes) and has high oil-generation potential.
- *Type II* is derived from marine organic matter consisting of animal and plant material. Its oil-generation potential is lower than that of type I but it is still a very important source of hydrocarbons. This is the most common kerogen worldwide. Middle East and northern North Sea source rocks contain type II kerogen.
- *Type III* is derived primarily from humic (terrestrial plant material) organic matter and is gas-prone. It is the dominant kerogen type in deltaic petroleum provinces.
- *Type IV* may have been derived from any source. It is usually recycled, oxidised and largely inert (very little oil and gas generation potential).

These different kerogen types follow different maturation pathways. The main changes in the chemical composition of kerogen with increasing maturity are reductions in their H/C and O/C ratios. Measurements of the H/C and O/C ratios are used in the *van Krevelen diagram* to (a) characterise the kerogen and (b) determine its level of maturity. Figure 4.17 presents a generic van Krevelen

Depth (km)

Relative quantity

Temperature (°C)

0

0

Original organic
chemicals

Diagenesis

Kerogen

Changes in molecular composition

3

75

$C_{34}H_{54}$

$C_{16}H_{18}$

Oil
window

Oil

$C_7H_{13}$

$C_3H_{16}$

6

150

Catagenesis

Natural
gas

Gas
window

$CH_4$

9

225

Metagenesis

Graphite

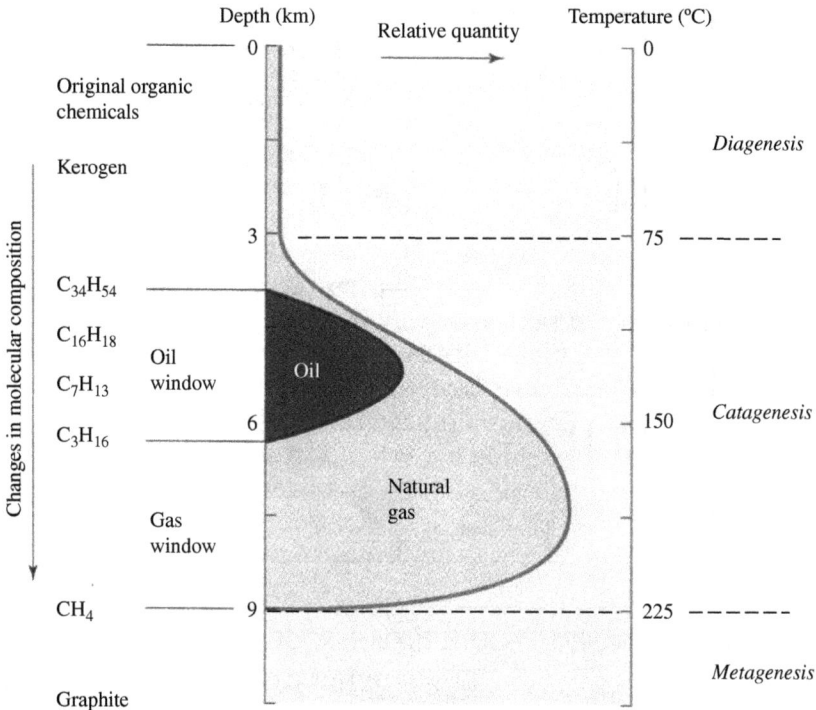

Figure 4.16. Formation of progressively lighter hydrocarbons with increasing temperature and depth (after Marshak, 2005).

diagram, indicating the general scheme of kerogen evolution and the principal products generated during each stage. *Diagenesis* refers to the changes occurring at relatively low temperatures while *catagenesis* is applied to the alterations taking place at elevated temperatures. Oil and gas are formed during the catagenetic phase. *Metagenesis* encompasses the changes when temperature rises above 225°C, resulting in the formation of graphite.

Figure 4.18 presents an example of a van Krevelen diagram for various kerogen types and their level of maturity.

Regional source-rock maturity studies are useful in outlining prospective areas for oil and gas field exploration. Figure 4.19 shows that the overwhelming majority of the hydrocarbon accumulations in the northern North Sea occur in the Viking and Central Grabens,

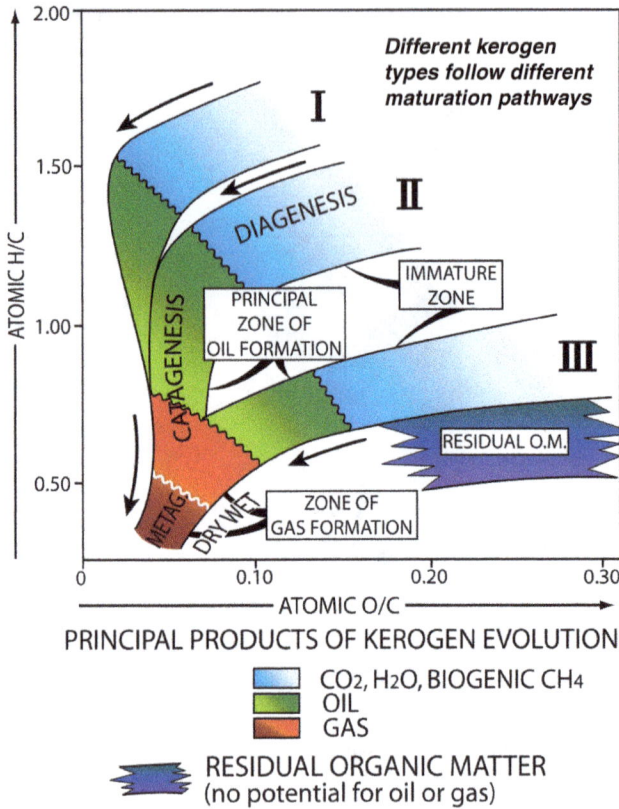

Figure 4.17. A generic van Krevelen diagram, showing the general scheme of kerogen evolution and the principal products generated during each stage.

where the Upper Jurassic Kimmeridge clay source rock has reached maturity. In these areas the drilling success ratio has been 1:3, but it drops to 1:30 where the source rock remains immature.

### 4.2.2.  *Estimating the time of hydrocarbon generation*

The van Krevelen diagram provides information on the type and level of maturity of kerogens but does not indicate the time of the onset of maturity and hence the start of hydrocarbon generation. The positions of the oil and gas windows in terms of time and depth in a given area can be determined by constructing *burial history* curves, which involve plotting depth of burial against geological time.

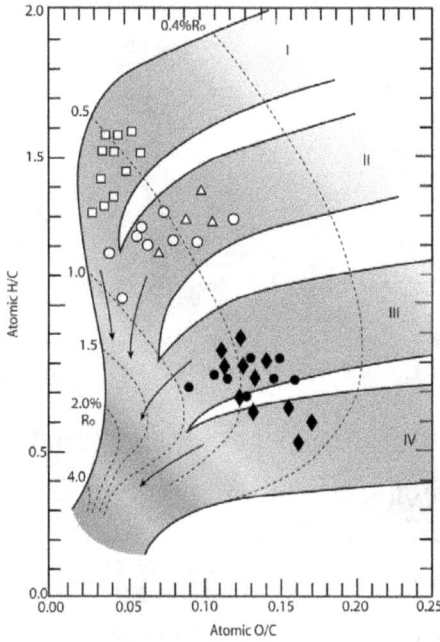

○ Type I kerogen, Eocene Green River Shale, USA      △ Type II kerogen, Jurassic shale, Saudi Arabia

○ Type II kerogen, Lower Jurassic shale, France      ● Type III kerogen, Tertiary shale, Greenland

◆ Type IV kerogen, Tertiary shale, Gulf of Alaska

Figure 4.18. A van Krevelen diagram for various kerogen types and their level of maturity (modified and redrawn from Hunt, 1995).

Figure 4.20 presents a burial history curve based on data from a well in the Gulf of Suez basin. The object of the exercise is to determine when the source rock, which is of Santonian (Upper Cretaceous) age and contains type II kerogen, became mature and entered the oil window.

The results show that maturity was reached and oil generation began in the Middle Miocene, about 9 million years ago, at a temperature of about 95°C and depth of 2,290 m. It ended in the Pliocene, about 2.5 million years ago, at a temperature of around 135°C. It should be noted, however, that the integration of time and temperature with depth is based on complex equations that involve calculations of a parameter known as the *time–temperature index*

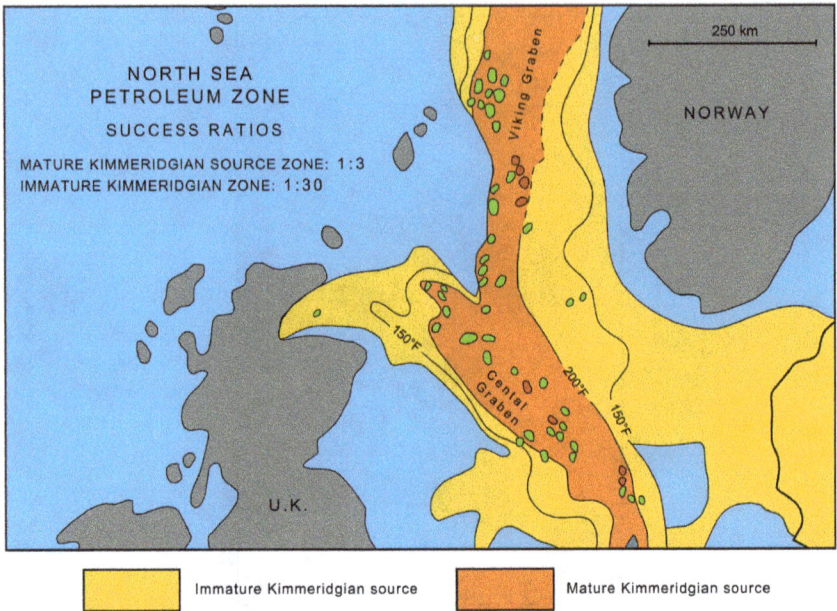

**NORTH SEA PETROLEUM ZONE**

**SUCCESS RATIOS**

MATURE KIMMERIDGIAN SOURCE ZONE: 1:3
IMMATURE KIMMERIDGIAN ZONE: 1:30

NORWAY

250 km

U.K.

☐ Immature Kimmeridgian source        ☐ Mature Kimmeridgian source

Figure 4.19. Map showing the association of the oil and gas fields with zones of source-rock maturity in the northern North Sea (modified and redrawn from Hunt, 1995).

(TTI), which is influenced not only by the thermal characteristics of a given area but also by the type of kerogen associated with the source rocks. In practice, the construction of burial curves is undertaken by using computer software specifically developed for this purpose. Comparing the time of oil and gas generation with the time of trap formation from geological studies is useful in determining whether the traps in a given area are likely to be hydrocarbon-bearing. It is essential that trap formation predate, or at least be simultaneous with, hydrocarbon generation and migration.

Finally, it should be mentioned that the ratio of reservoired oil to petroleum generated in-place in a typical basin is estimated to be less than 15%; therefore, a great proportion of the hydrocarbons generated in the source rocks remains therein. This underpins the unconventional shale oil and shale gas production industries — extraction of the hydrocarbons that have remained in the shale source rocks.

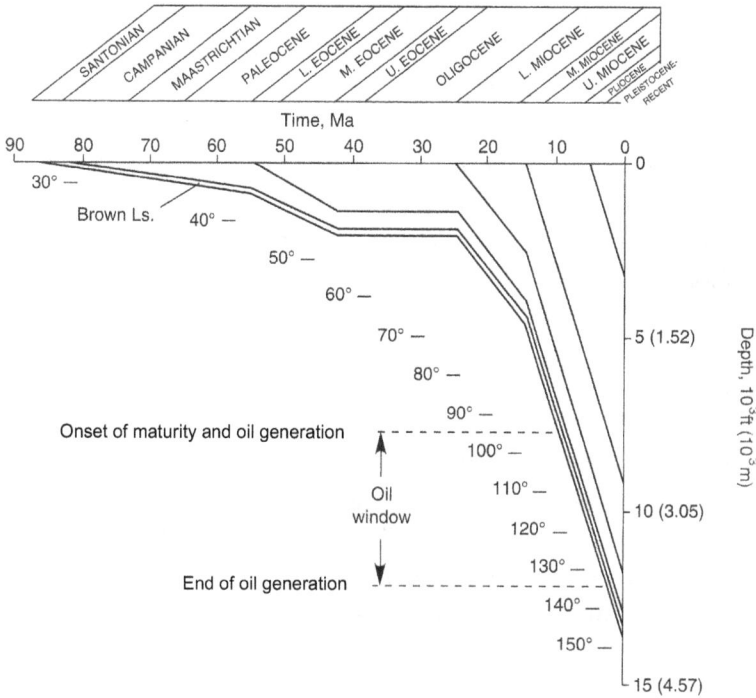

Figure 4.20. Burial history curve based on data from a well for the source rock in the Gulf of Suez basin (modified and redrawn from Hunt, 1995).

## 4.3. Migration of Petroleum

Once generated, the hydrocarbons are expelled from the source rock. It is generally agreed that migration is triggered by the natural compaction of the source rock and by the process of oil and gas generation. Fine-grained sediments initially have a high porosity and contain large volumes of water. With increasing overburden pressure resulting from burial, the water is squeezed out together with any generated hydrocarbons. If the water cannot escape fast enough, as is often the case from fine-grained sediments, it becomes at least partially trapped, causing the pressure to build up in the source rock, resulting in over-pressuring. Moreover, as the oil and gas separate from the kerogen during generation, they take up more space and further increase the pressure in the source rock, thus helping to maintain its permeability.

It is now recognised that over-pressuring is a feature associated with source rocks. As already noted, over-pressuring helps preserve the permeability of shales, thereby facilitating the movement of fluids. The movement of the hydrocarbons out of the source rock into the reservoir is called *primary migration*. The movement can be upward, lateral or even downward, depending on the relative position of the source and reservoir rocks. The actual mechanism of primary migration has been a matter for debate for decades. It is not known whether the hydrocarbons move out of the source rock in solution, in colloidal suspension, as "protopetroleum" (something that changes into oil and gas after entering the reservoir), as discreet globules of oil or by diffusion.

The movement of oil and gas within the reservoir is known as *secondary migration*. Secondary migration results in accumulation, i.e. the formation of oil and gas fields. It is generally accepted that the main cause of secondary migration is the buoyancy of oil and gas. They are both lighter than water and their distribution in the reservoir is in accordance to their densities; being the lightest fluid, gas moves to the top in the trap and is underlain by oil, which lies above the water bearing zone.

Migration is a slow process, with oil and gas travelling between hundreds of metres and several kilometres over millions of years to charge traps. There is thus no question of natural processes replacing the reserves that are currently being depleted.

## 4.4. Reservoir Rocks

In conventional oil and gas fields, hydrocarbons are stored in reservoir rocks. To function as a reservoir, the rock must possess certain fundamental properties, namely porosity and permeability.

Although fractured shale and weathered igneous and metamorphic rocks function as reservoirs in some instances, an example of which is presented in Fig. 4.21, the overwhelming majority of oil and gas fields worldwide are housed in sandstones and carbonates. The latter include limestones, dolomites and chalk.

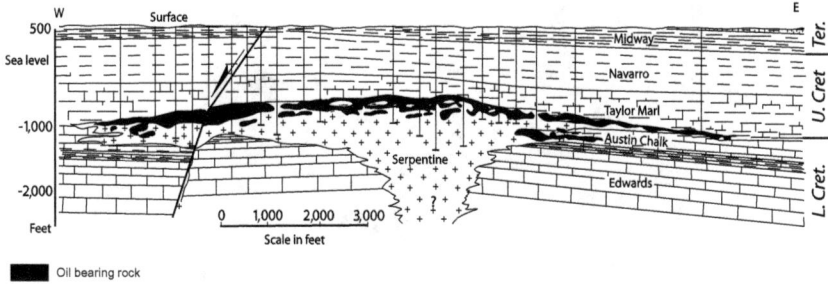

Oil bearing rock

Figure 4.21. The Lytton Springs oil field, Texas. The reservoir is a serpentine, an altered igneous rock. Exposure to surface conditions has resulted in weathering and fracturing has created porosity and permeability, enabling rock to function as a reservoir (modified and redrawn from Levorsen, 1967).

### 4.4.1. *Porosity*

Usually abbreviated to $\Phi$ and expressed as a percentage, *porosity* is a measure of the storage capacity of the reservoir. It is defined as

$$\Phi = (\text{Pore volume})/(\text{Bulk volume}) \times 100. \qquad (4.1)$$

In sandstones, the pores or voids usually occur between the grains that make up the rock. This is referred to as *intergranular* or *primary* porosity and represents an inherited or depositional feature; i.e. it developed at the time of the accumulation of the sand grains. In carbonates the voids usually, but not always, occur between the crystals that form the rock. This type of porosity is called *intercrystalline* porosity. However, carbonate rocks are susceptible to diagenetic changes after deposition and consolidation. These changes include solution and fracturing, both of which can increase porosity. This type of porosity is referred to as *secondary* porosity. Pores of this origin can be quite large and are known as *vugs*.

Figure 4.22 shows a thin section of a porous sandstone and Figs. 4.23 and 4.24 provide examples of porous carbonates. A photograph of a limestone with vugs is presented in Fig. 4.25.

The most common porosity range is 10–20% and the highest porosity value recorded in the literature is 37%. The maximum

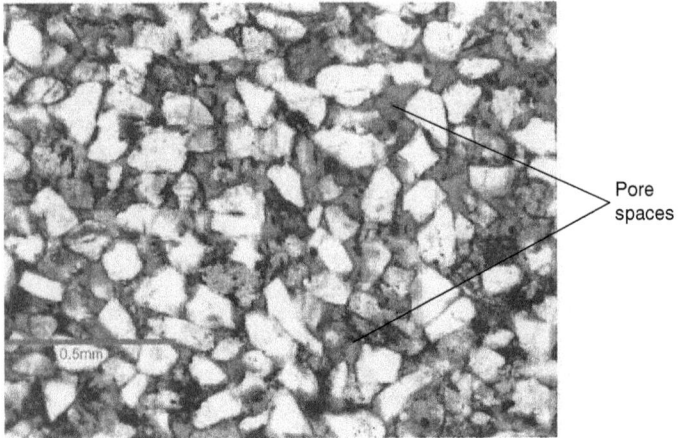

Figure 4.22. A thin section of a porous sandstone (after Aqrawi *et al.*, 2010).

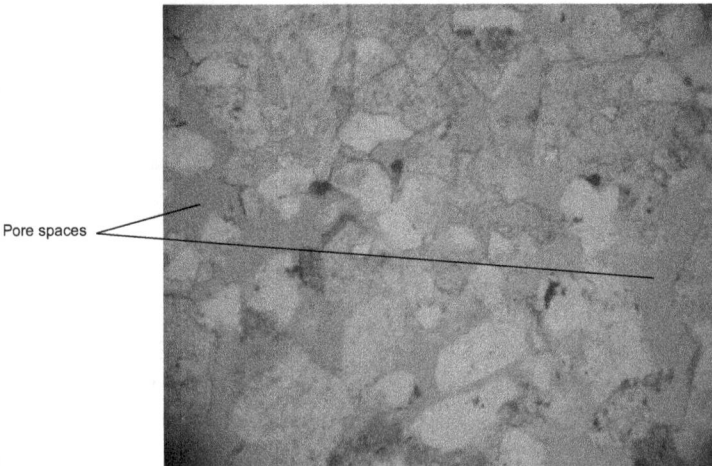

Figure 4.23. A thin section of a porous limestone (http://www.churchmonu mentssociety.org/images/Geology/img1.gif).

theoretical porosity value is 47% (based on the grains being perfect spheres and loosely packed). In the case of clastic rocks, i.e. sandstones, porosity tends to decrease with increasing depth of burial (due to compaction) and increasing age. As shown in Fig. 4.26, for the

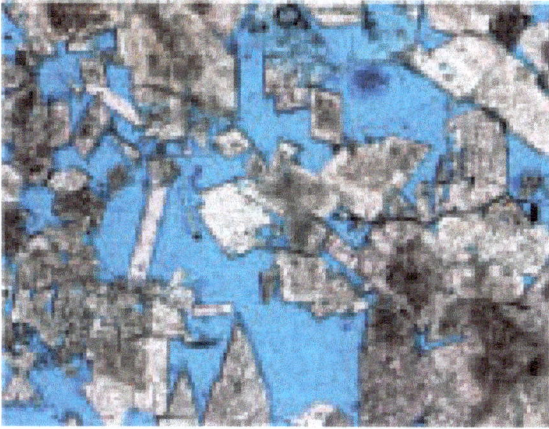

Figure 4.24. A thin section of a porous dolomite (after Lindsay *et al.*, 2006).

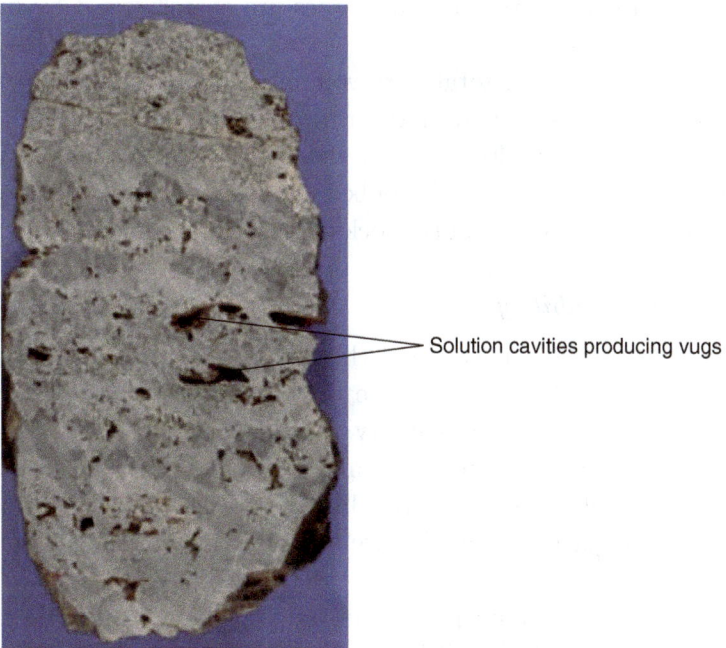

Solution cavities producing vugs

Figure 4.25. Vuggy porosity in a limestone hand specimen (photo by M. Ala).

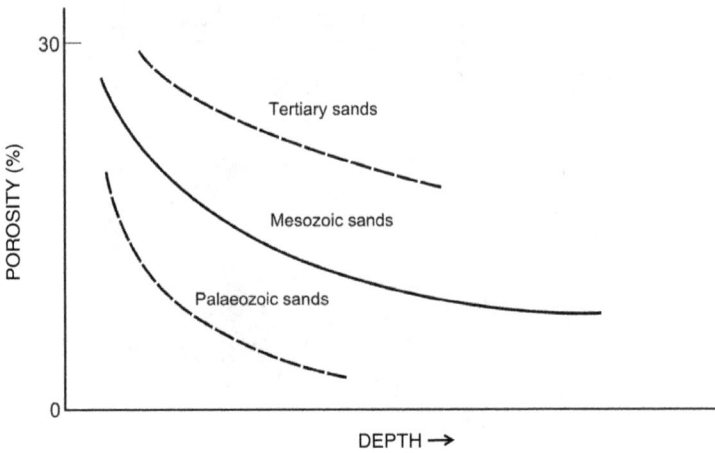

Figure 4.26. Graph demonstrating reduction in porosity with increasing depth of burial and age (modified and redrawn from North, 1985).

same burial depth, younger sandstones are characterised by higher porosity.

The ratio between total pore volume and bulk volume is called the *absolute* porosity, $\Phi_A$, of the rock. What is more important is its *effective* porosity, $\Phi_E$, which is defined as the ratio between the *interconnected* pore spaces and bulk volume. $\Phi_E$ is usually lower than $\Phi_A$ and the permeability of the rock depends on its effective porosity.

### 4.4.2.   *Permeability*

*Permeability* $(k)$ is a measure of the ability of the rock to transmit fluids and depends on the degree of connection between the pore spaces, i.e. on $\Phi_E$. Permeability is a complex quantity and is influenced by several factors, the most important of which are fluid saturation, fluid viscosity and the diameter of the passages or throats connecting adjacent pores. Figure 4.27 provides an illustration of permeability.

When only one fluid is present and it fully saturates the rock, the permeability of the rock to that fluid is a maximum and is called its *absolute* permeability, $k_A$. When more than one fluid is

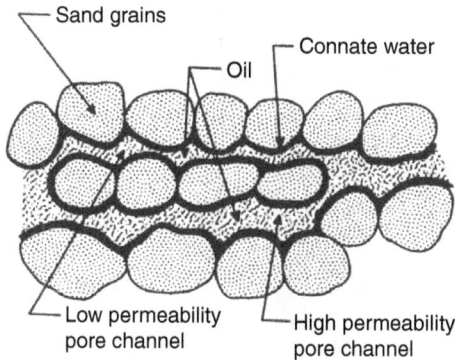

Figure 4.27. Illustration of permeability (from Clark, 1969).

present, as is the case in most reservoirs, permeability to any one fluid is reduced. The ability of a reservoir to conduct one fluid in the presence of others depends on the saturation of that fluid and is called its *effective* permeability, $k_E$. $k_E$ changes as the saturation of the fluid in question varies. In reservoir studies, a quantity known as *relative* permeability, $k_r$, is used and defined as the $k_E/k_A$ ratio. With only one fluid present, $k_E = k_A$ which means that $k_r = 1$. Its value therefore ranges from zero to one (depending on fluid saturation), which makes it easier to handle in reservoir studies.

The distinction between the "high-permeability" and "low-permeability" channels in Fig. 4.27 is based on the differences between the diameters of the passages connecting adjacent pores. The high-permeability channel is characterised by larger-diameter pore throats that make it easier for the fluids to pass through the rock. Oil and gas flow rates from producing wells increase with increasing permeability of the reservoir. The unit of permeability is the darcy, named after Henry Darcy, the French engineer who defined the parameters governing the flow of water through porous media in the 1850s. In practice, the darcy is too large since the permeability of reservoirs is usually below this level. Accordingly, permeability is recorded in millidarcies, $1/1,000^{\text{th}}$ of a darcy, abbreviated to md or Md.

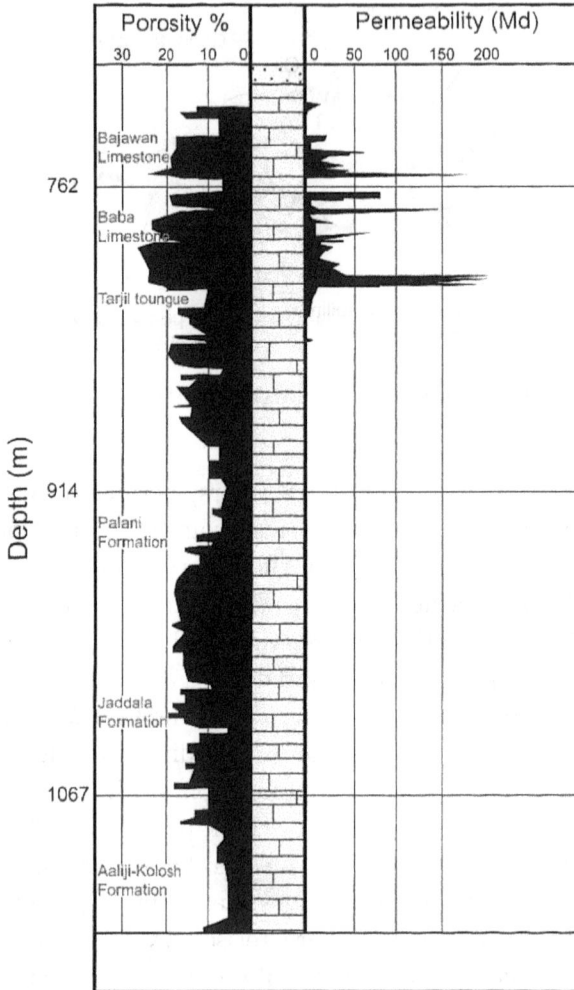

Figure 4.28. Poor correlation between $\Phi - k$ distribution in the lower 2/3 of a section through the carbonate reservoir in the Kirkuk Field, northeastern Iraq (after Jassim and Al-Gailani, 2006, courtesy M.B. Al-Gailani).

### 4.4.3.  *Porosity–permeability relationship*

Although high permeability values often correspond to high porosity values, the corollary of this statement is not always true; i.e. high porosity zones are not necessarily characterised by high permeability.

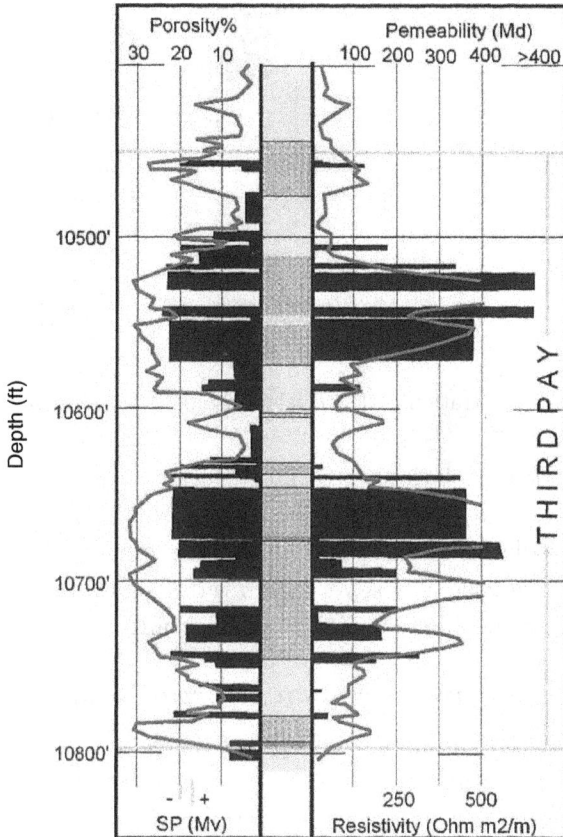

Figure 4.29. Good match between $\Phi-k$ distribution in part of the clastic reservoir in the Zubair Field, southern Iraq (after Jassim and Al-Gailani, 2006, courtesy M.B. Al-Gailani).

It should be noted that this is more common in carbonate than in clastic reservoirs, as illustrated in Figs. 4.28 and 4.29. There is no correlation between the distribution of porosity and permeability in the lower two-thirds of the well section from the carbonate reservoir in the giant Kirkuk Oil Field in northeast Iraq, as shown in Fig. 4.28. By contrast, in the well profile in Fig. 4.29, depicting part of the clastic reservoir in the Zubair Field in southern Iraq, there is a good match between the porous and permeable intervals.

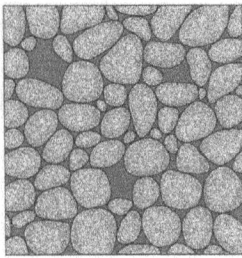

Low porosity, low permeability due to poor connectivity between adjacent pores. Result: low$\Phi_A$, low $\Phi_E$.

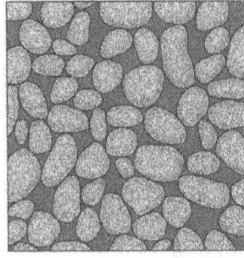

Porosity is high in this case, but permeability remains low, since connectivity between adjacent pores is still poor. Result: high $\Phi_A$, low $\Phi_E$.

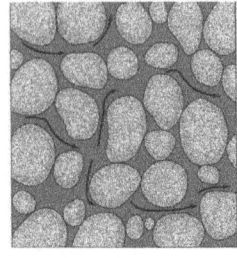

High porosity and high permeability due to good vertical and lateral connectivity between the pores. Result: high $\Phi_A$, high $\Phi_E$.

Figure 4.30. Illustration of $\Phi - k$ relationships (after Marshak, 2001).

Poor $\Phi - k$ correlations invariably result from large differences between $\Phi_A$ and $\Phi_E$, as explained in Fig. 4.30. A thin section of a rock with isolated pore spaces and tight matrix is presented in Fig. 4.31. In this case, $\Phi_A$ is very high but $\Phi_E$ is zero, which means that the rock would have negligible permeability.

### 4.4.4.  *Effects of fracturing on reservoir properties*

*Fracturing* greatly enhances permeability. It results from shrinkage as the sediment loses water (diagenetic fractures) or tectonic movements (tectonic fractures) and develops in three directions, mutually at right angles to one another, as illustrated in Fig. 4.32. Figure 4.33 shows the outcrop of a fractured limestone of Jurassic age in the Bristol Channel, southwest England. Diagenetic fractures develop early, before the rock is fully indurated and in a semi-brittle state, while tectonic fractures form as a result of the folding of brittle rock layers. Fracturing is particularly important in carbonate reservoirs. The early diagenetic fractures are widened by later folding and this greatly enhances strike-parallel permeability, as shown diagrammatically in Fig. 4.34.

The best known examples are in the Middle East, where individual well flow rates of up to 100,000 b/d have been recorded from fractured limestone reservoirs with relatively modest matrix porosity (9–14%) and permeability (tens of Md). Consequently, in addition

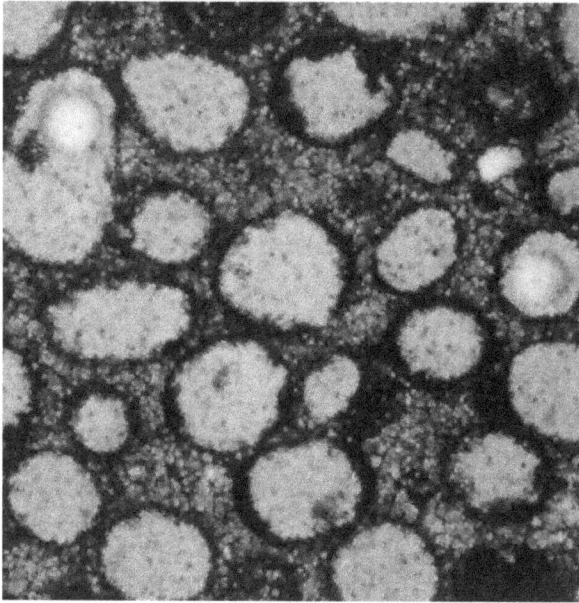

Figure 4.31. A rock with isolated pore spaces and tight matrix. High $\Phi_A$ but virtually no $\Phi_E$ results in very low or no permeability (from Selwood, 2007).

Figure 4.32. Occurrence of fractures in a rock (after Marshak, 2005).

Figure 4.33. Fractured Jurassic limestone, Bristol Channel, southwest England (photo by J. Cosgrove).

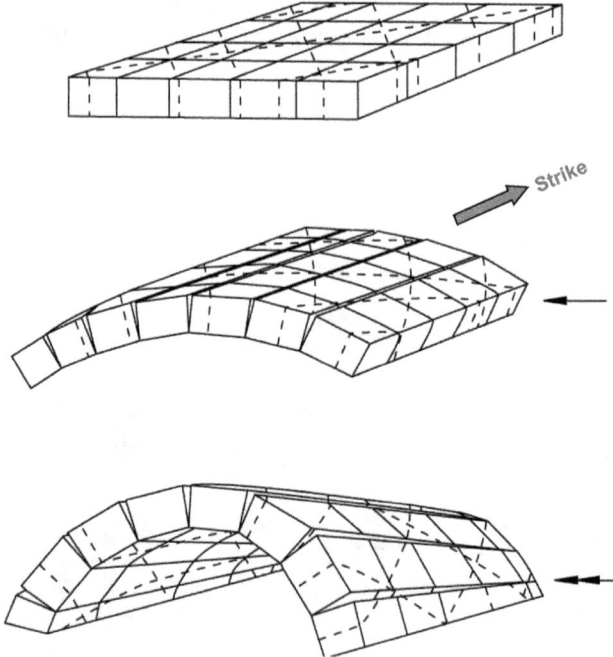

Figure 4.34. Folding widens the early diagenetic fractures, enhancing strike-parallel permeability (modified and redrawn from McQuillan, 1973).

to matrix porosity and permeability, fractured reservoirs also have fracture porosity and permeability. Such reservoirs are called *dual porosity/permeability systems.*

## 4.5.   Traps

A *trap* is a special situation in the reservoir that arrests the migration process and causes the hydrocarbons to accumulate. Traps may be formed by changes in the shape of the reservoir (structural traps) or by variations in its lateral continuity (stratigraphic traps) or by a combination of these factors (combination traps). Trap formation must predate hydrocarbon migration.

A simple trap classification scheme is presented in Table 4.1 and Fig. 4.35 provides a diagrammatic illustration of the various trap types.

It should be noted that salt domes appear in all three categories.

### 4.5.1.   *Structural traps*

*Structural traps* include anticlines and fault- and thrust-associated features. Globally, anticlines are the most common traps and an overwhelming majority of the giant fields (accumulations containing

Table 4.1.  A simple trap classification scheme.

| Category | Association |
| --- | --- |
| Structural | Folds |
| | Salt domes and other intrusions |
| | Horsts |
| | Faults and thrusts |
| | Buried hills and reefs |
| Stratigraphic | Depositional changes |
| | Salt domes |
| | Reefs |
| | Unconformities |
| | Asphalt clogging/cementation |
| Combination | Up dip pinch out |
| | Salt dome |

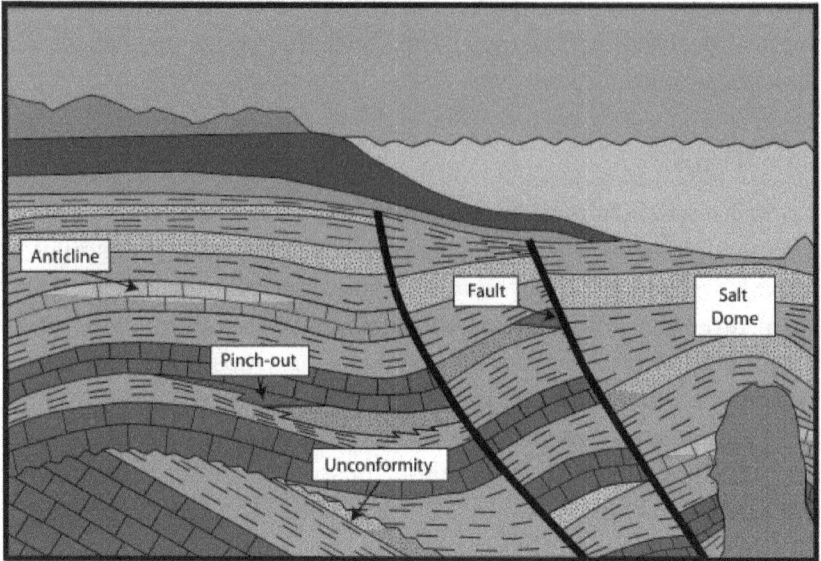

Figure 4.35. Diagrammatic illustration of hydrocarbon trap types.

reserves exceeding 500 million barrels of oil or gas equivalent) are of this type. A typical example of an anticlinal trap is illustrated in Fig. 4.36 and the elements of an anticlinal trap and the associated terminology are explained in Fig. 4.37.

*Closure* is the vertical distance between the crest of the structure and spill point. It is an important feature of an anticlinal trap: It controls the height of the hydrocarbon column and hence the storage capacity of the structure. *Spill point* is defined as the lowest level at which hydrocarbons can be retained in the trap. Once a trap has been filled to its spill point, its hydrocarbon storage capacity ends; any more hydrocarbons moving into the trap will spill out and continue to migrate until they encounter another trap along their migration path. The fluid contacts are normally horizontal, but tilted contacts are known to occur due to a variety of factors. Perhaps the most common cause is the flow of water through the reservoir, which exerts a drag on the fluid contacts, displacing the hydrocarbons downflank as illustrated in Fig. 4.38. The tilt is in the direction of

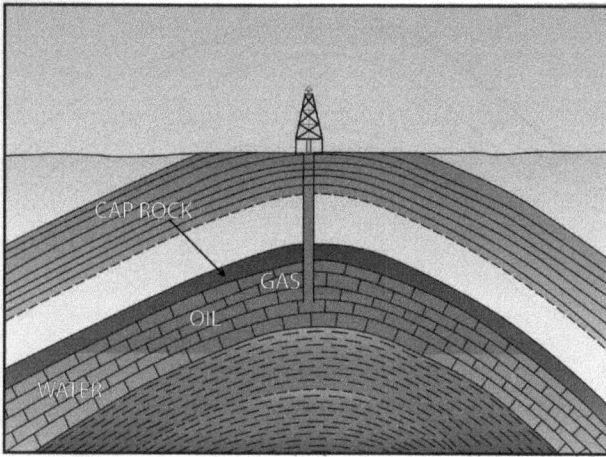

Figure 4.36. A typical anticlinal trap.

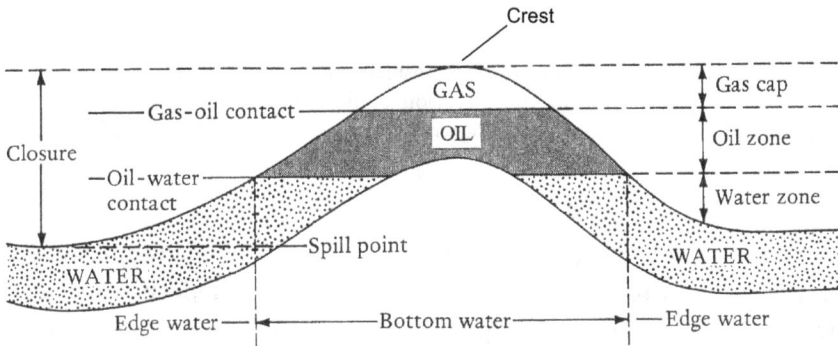

Figure 4.37. Elements of an anticlinal trap (after Selley, 1997).

flow and the degree of down dip displacement of the hydrocarbons depends on the strength of the water flow.

Anticlinal traps can be produced by compressional and non-compressional processes. Horizontal forces are involved in the formation of compressional anticlines, while vertical movements, often associated with faults, are responsible for the production of non-compressional anticlines. Irregularities in the basin floor in the form of hills can also create anticlines by causing the upward arching of the

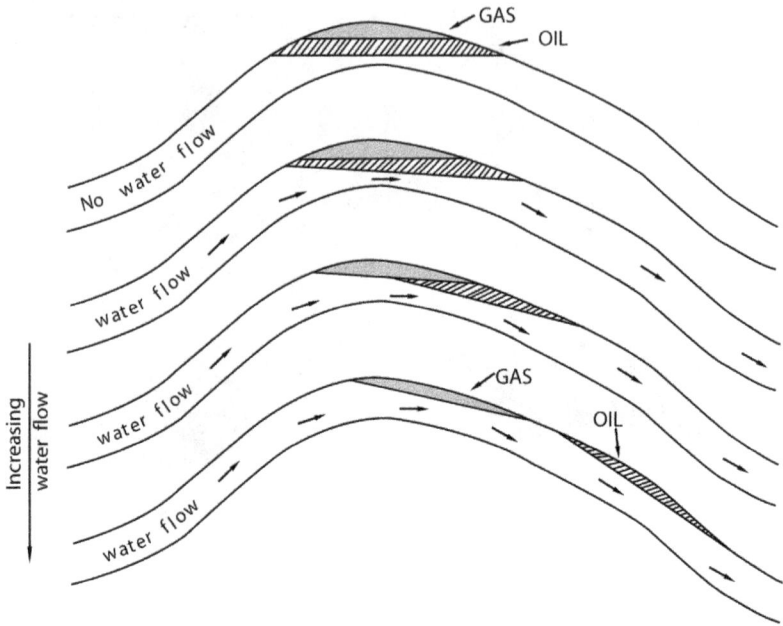

Figure 4.38. Tilted oil–water contacts resulting from the flow of water through the reservoir. The tilt is in the direction of flow and the degree of down dip displacement of the hydrocarbons depends on the strength of the waterflow (after Hobson and Tiratsoo, 1985).

strata deposited above such features, as illustrated diagrammatically in Fig 4.39. Moreover, there is a difference in depositional rates over the top of the hills compared to the flanks, resulting in greater sedimentary thicknesses in the latter areas. The degree of compaction is greater for the thicker flank sediments, causing them to subside more, and this enhances the amplitude of the anticline, which in turn increases the closure and hence the storage capacity of the trap. Structures of this type are referred to as drape or compaction anticlines and there are several hydrocarbon accumulations associated with such features in the North Sea, including the Forties Field (Fig. 4.40).

Perhaps the best-known examples of compressional anticlinal traps are located in the Zagros Mountain foothills of the Middle East Basin in southwest Iran and their continuation into northern Iraq.

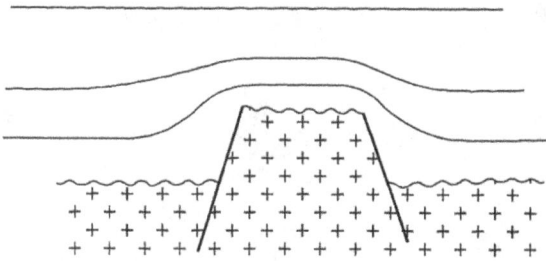

Figure 4.39. Diagrammatic illustration of the formation of an anticline by draping over an irregularity in the basin floor (after Allen and Allen, 2005).

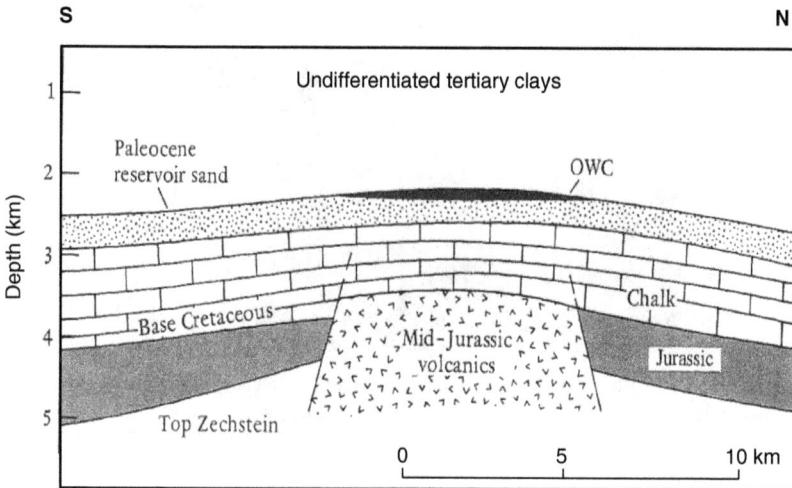

Figure 4.40. The Forties Field, North Sea. This is an example of a drape anticline over a basement high (after Selley, 1997).

This is a belt of generally northwest–southeast trending elongated anticlines of great length and amplitude that host supergiant oil and gas accumulations. Individual structures reach lengths of up to 100 mi and closures of about 10,000 ft. They were formed by strong northeast–southwest compressional forces and are clearly visible in the satellite image of the southeastern part of the Zagros fold belt in southeast Iran in Fig. 4.41. A northeast–southwest cross-section through the Iranian Zagros sector is shown in Fig. 4.42. The

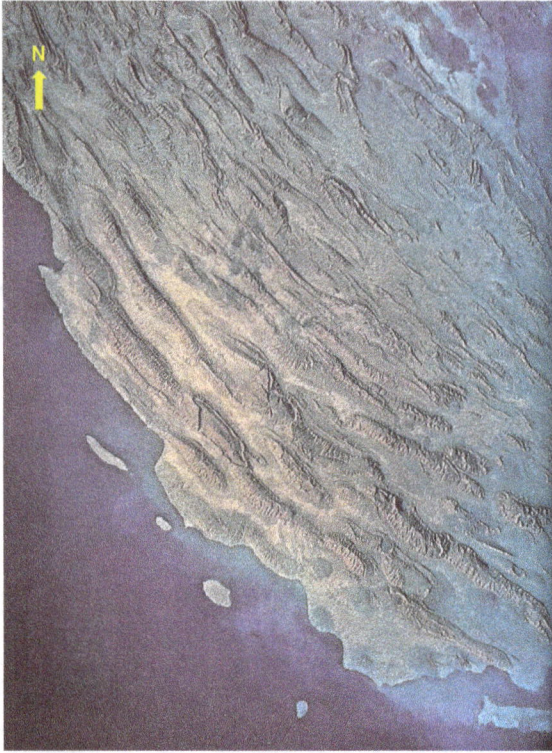

Figure 4.41. Satellite image of the southeastern part of the Zagros fold belt, southwest Iran (formerly at http://en.wikipedia.org/wiki/Persian_Gulf).

anticlines are characterised by large amplitudes and many of them are asymmetrical with thrust-faulted southwestern limbs in some cases.

Non-compressional anticlines differ in terms of their structural style from their compressional counterparts. The former tend to be symmetrical and have low-amplitude features as shown in Fig. 4.40. Table 4.2 provides a comparison between compressional and non-compressional structural styles and their associated features.

Faults and fault-induced features can also form traps. Such traps may be associated with both normal and reverse faults and Fig. 4.43 shows the essential requirements for faults to form traps. The main condition is that the permeable layer must be juxtaposed across the fault against an impermeable bed to cause accumulation. Unless

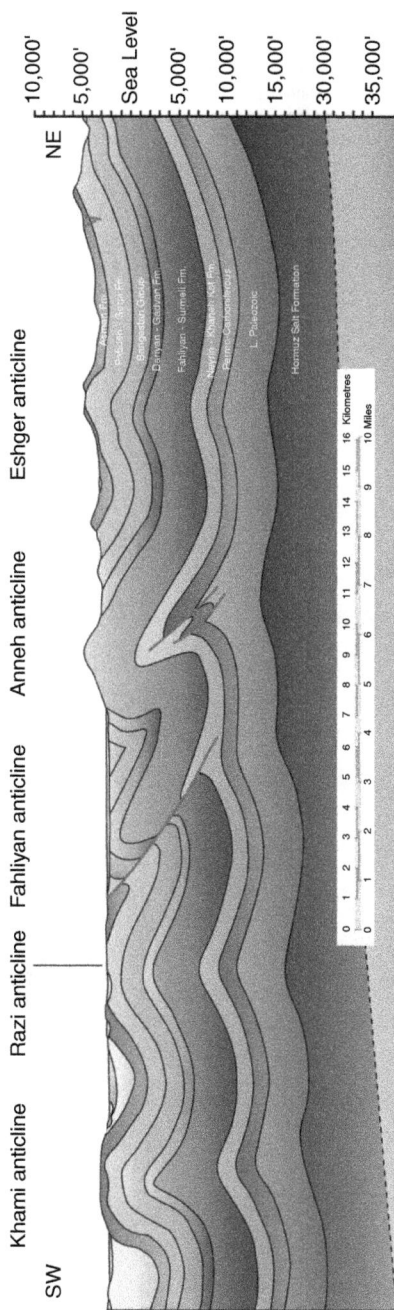

Figure 4.42. Cross-section showing structures associated with compressional movements in the Zagros Basin, southwest Iran (Oil Service Company of Iran, 1974).

Table 4.2. A comparison of compressional and non-compressional structural styles.

|  | Compressional | Non-compressional |
|---|---|---|
| FOLDS | Large amplitude, may be asymmetrical, one limb may be thrust faulted | Low amplitude, generally symmetrical, normally faulted limbs |
| FAULTS | Reverse, transcurrent and some normal faulting, thrusting common | Largely normal faulting, thrusts generally absent |
| FRACTURES | Largely due to buckling | Largely due to shrinkage and other diagenetic factors |
| CRUSTAL RESPONSE | Shortening | Extension |

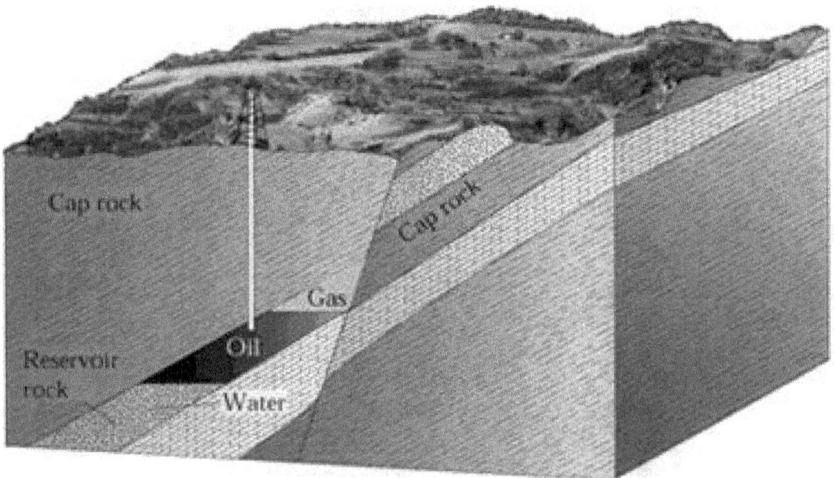

Figure 4.43. Elements of a fault trap (http://www.mpgpetroleum.com/images/fault.jpg).

there is a barrier to arrest hydrocarbon movement, oil and gas continue to migrate and no accumulation will be formed. The fault plane must also be sealing to prevent leakage and dissipation of the hydrocarbons and, ideally, the throw of the fault should be greater than reservoir thickness. Figure 4.44 illustrates the impact of the

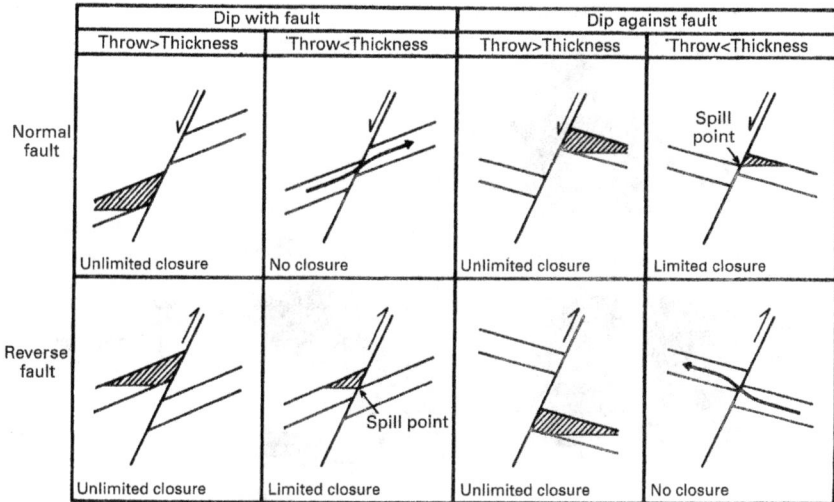

| | Dip with fault | | Dip against fault | |
| --- | --- | --- | --- | --- |
| | Throw>Thickness | Throw<Thickness | Throw>Thickness | Throw<Thickness |
| Normal fault | Unlimited closure | No closure | Unlimited closure | Spill point — Limited closure |
| Reverse fault | Unlimited closure | Spill point — Limited closure | Unlimited closure | No closure |

Figure 4.44. Closure associated with fault-controlled traps. The amount of closure depends on (a) the dip of the beds and (b) the throw of the fault in relation to reservoir bed thickness (after Stoneley, 1995).

throw of the fault in relation to reservoir thickness on whether or not an effective trap is present. Unlimited closure results from the throw of the fault exceeding reservoir thickness but the integrity of the trap becomes compromised if the throw of the fault is less than that of the permeable layer. In the latter case, the reservoir remains in contact with its extension across the fault plane, which means that there is no barrier to hydrocarbon migration, leading to limited or no closure.

Many of the oil and gas fields in the North Sea and the Gulf of Suez are associated with horsts and tilted fault blocks. Figure 4.45 presents a cross-section of the Gawain Gas Field, a horst-block-associated trap in the southern North Sea. Examples of tilted-fault-block-controlled traps in the northern North Sea and the Gulf of Suez are shown in Figs. 4.46 and 4.47, respectively.

Thrusting occurs in regions that have been subjected to strong horizontal compression. These regions include mountain belts where thrust-associated traps have been found above as well as below thrust planes. Traps of this type are common along the eastern flank of the

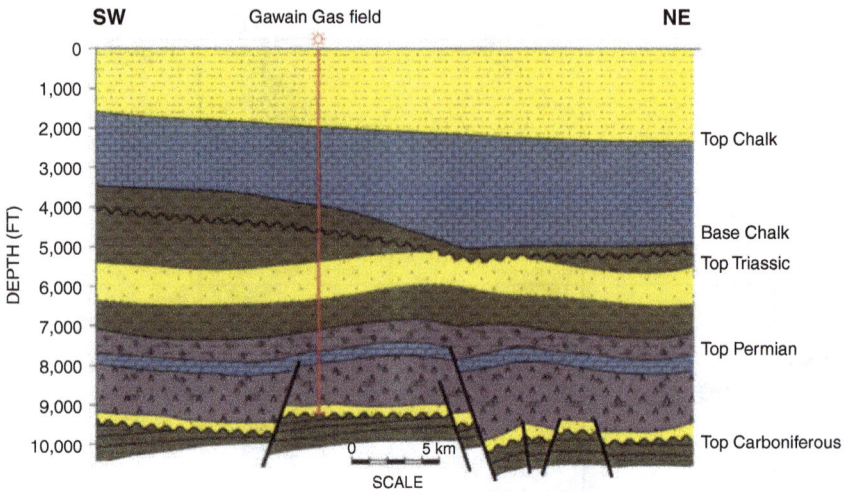

Figure 4.45. Cross-section of the Gawain Gas Field, southern North Sea, an example of a trap associated with horst blocks (after Osbon *et al.*, 2003).

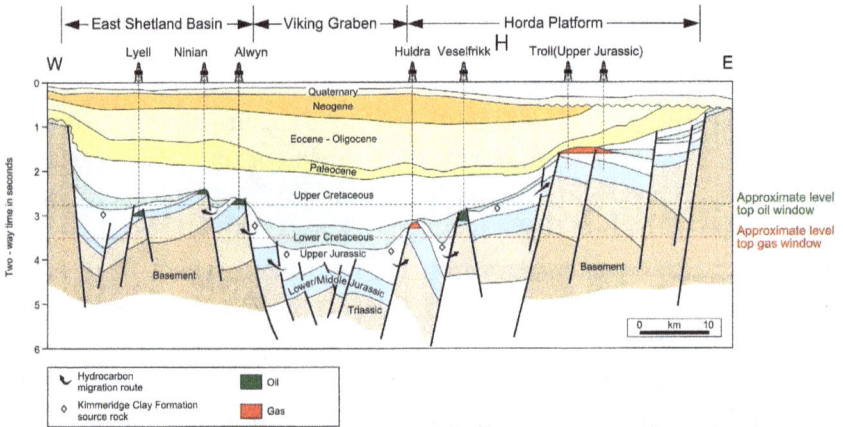

Figure 4.46. A diagrammatic section across the northern North Sea showing hydrocarbon accumulations associated with tilted fault blocks (after Underhill, 2003).

Rocky Mountains in Canada and the United States and Fig. 4.48 presents a cross section of the Kavik Gas Field in western Canada, which serves as a good example of a thrust-associated trap (see also Chapter 1, Fig. 1.78).

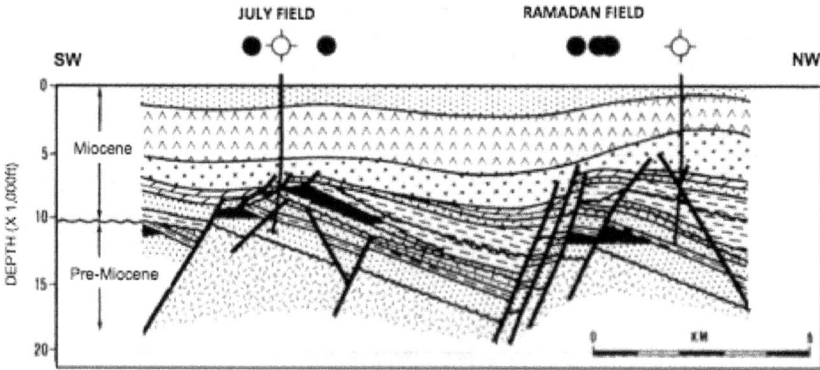

Figure 4.47. Cross-section through the July and Ramadan Oil Fields, Gulf of Suez. These are examples of tilted-fault-controlled traps (after Clifford, 1986).

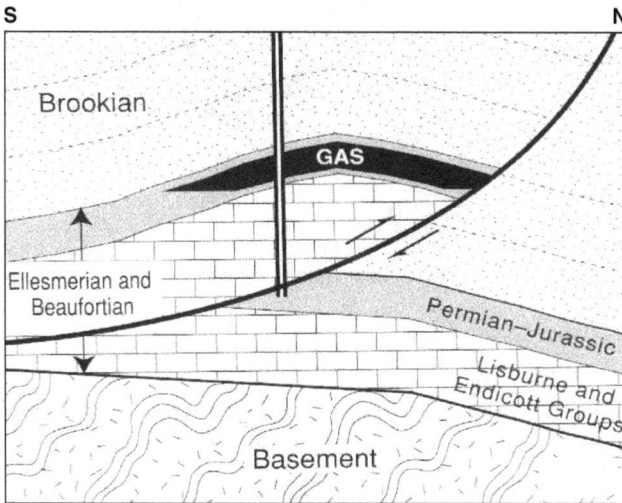

Figure 4.48. Cross-section of the Kavik Gas Field, western Canada, a thrust-associated trap (after Bird, 2001).

## 4.5.2. *Stratigraphic traps*

*Stratigraphic traps* are created by variations in lithology and, other than gentle tilting, are largely independent of structural deformation. They are more subtle and more difficult to identify in the sub-surface than structural traps and are formed by a variety of geological

Table 4.3. A simple classification of stratigraphic traps.

| | |
|---|---|
| Depositional | Pinch outs |
| | Lateral facies changes |
| | Reefs |
| Diagenetic | Porosity/permeability reduction due to cementation |
| Unconformity related | Above the unconformity surface |
| | Below the unconformity surface (truncation) |

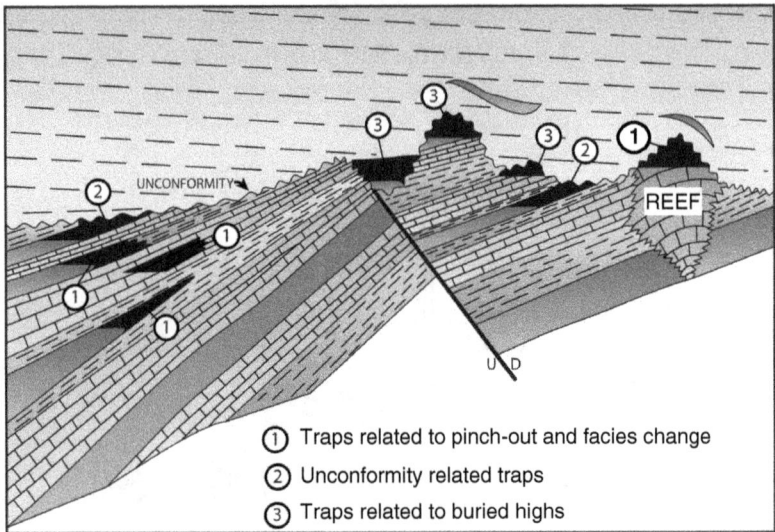

Figure 4.49. Illustration of the various types of stratigraphic traps (modified from Halbouty, 1972).

phenomena, including depositional, diagenetic and unconformity-related features. A simple classification scheme presented in Table 4.3 and Fig. 4.49 provides a diagrammatic illustration of the various stratigraphic trap types.

An example of an oil field producing from a pinch-out-type stratigraphic trap is the Port Acres Oil Field, Texas. As shown in Fig. 4.50, the producing sandstone pinches out in an east–west direction.

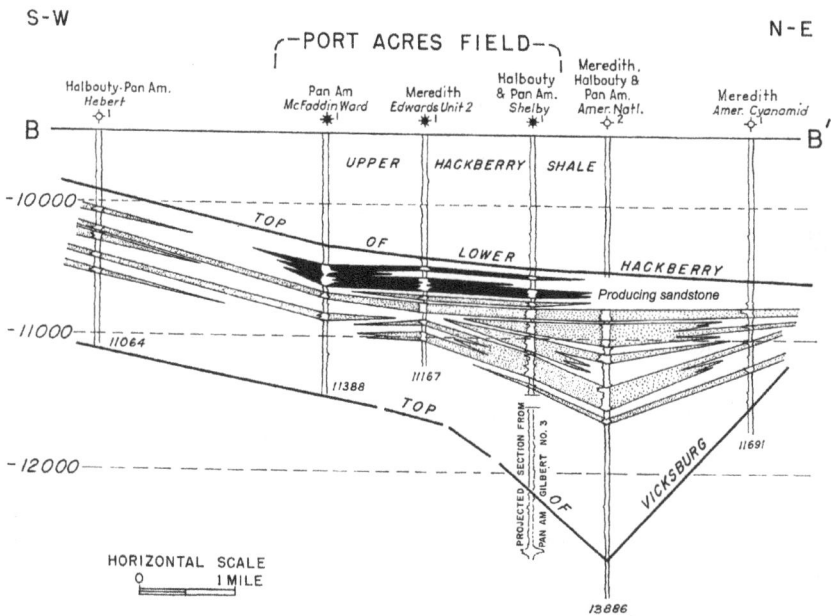

Figure 4.50. Cross-section of the Port Acres Oil Field, Texas, a trap associated with a sandstone pinch-out. The producing sandstone pinches out in an east–west direction (after Halbouty and Barber, 1972).

Several reef-associated accumulations are known from the Sirte basin of Libya, with the reefs having developed over horst blocks as depicted in Fig. 4.51. The best-known example of these is the giant Intisar Oil Field, producing from a circular reef of Tertiary age.

Figure 4.52 presents an east–west cross-section through the East Texas Basin, showing the location of the East Texas Oil Field. This is a classic example of an accumulation associated with a sub-unconformity trap. It was a giant field, discovered in the late 1920s, and produced more than 5 billion barrels of oil before being abandoned in the 1970s.

### 4.5.3. *Combination traps*

The distinction between structural and stratigraphic trapping is not always clear-cut and in many fields a combination of these factors has played a vital role in the creation of the trap. Figure 4.53 illustrates the elements of a combination trap; the pinch-out of the

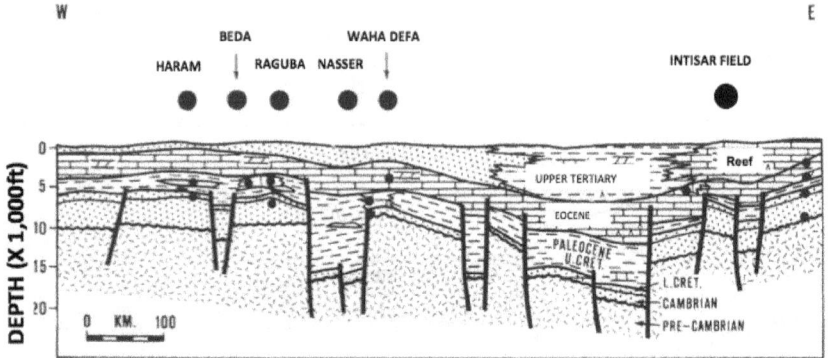

Figure 4.51. Diagrammatic cross-section through the Sirte basin, Libya. The Intisar Oil Field is a giant oil accumulation associated with a reef (after Clifford, 1986).

sand above the unconformity provides the stratigraphic element and tilting represents the structural aspect. The structural element is obvious and its identification most probably led to the discovery of the accumulation.

The giant Prudhoe Bay Oil and Gas Field, Alaska, is a good example of a combination trap. As shown in Fig. 4.54, it is a broad, east–west-trending, fault-bound anticline, the identification of which by seismic surveying led to the discovery of the field in 1968. Rocks ranging in age from the Carboniferous to the Lower Cretaceous were folded into an anticline and tilted towards the west at the end of the Early Cretaceous, resulting in truncation by an easterly dipping unconformity. They were subsequently overlapped by Cretaceous shales, which provide the seal to the Triassic sandstone reservoir.

An east–west section across the Prudhoe Bay Field is presented in Fig. 4.55 and shows it to be an unconformity- and fault-associated combination trap.

In conclusion, the following aspects of combination traps are noteworthy:

- it is the structural element that usually leads to their discovery;
- the hydrocarbons must have migrated into the traps after the deposition of the overlying seal to prevent their loss.

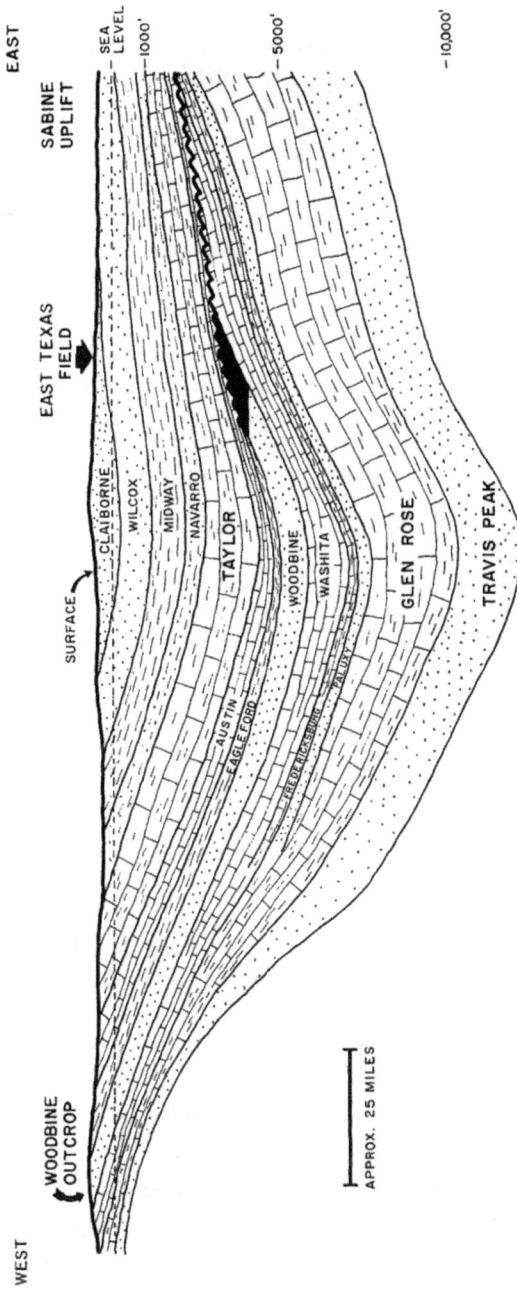

Figure 4.52. Diagrammatic east–west cross-section through the East Texas basin, showing the location of the East Texas oil field, a giant accumulation associated with a sub-unconformity trap (after Halbouty, 1972).

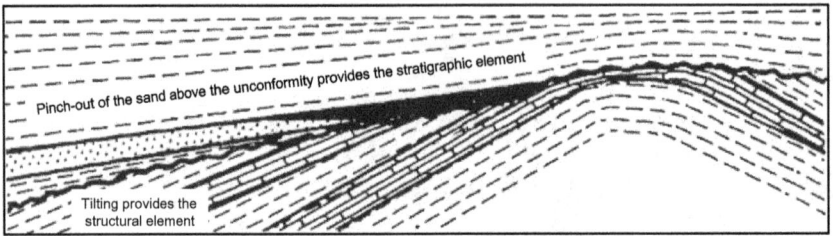

Figure 4.53. Diagrammatic illustration of a combination trap (after Rittenhouse, 1972).

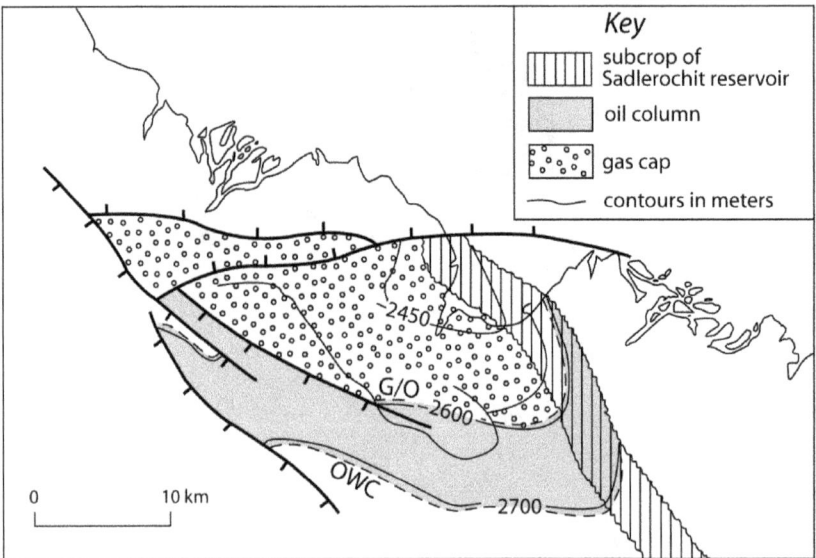

Figure 4.54. Outline of the Prudhoe Bay Field, showing the oil- and gas-bearing parts of the reservoir (modified and redrawn from North, 1985).

### 4.5.4. *Traps associated with salt movement*

As mentioned in Chapter 1, evaporitic rocks behave in a plastic manner and flow when subjected to stress. The flow takes place from regions of high stress to those of low stress and once triggered, the movement gives rise to salt-induced features, which include salt pillows or turtlebacks, piercement domes (also referred to as plugs or diapirs) and salt walls. The causes of the rise of salt through

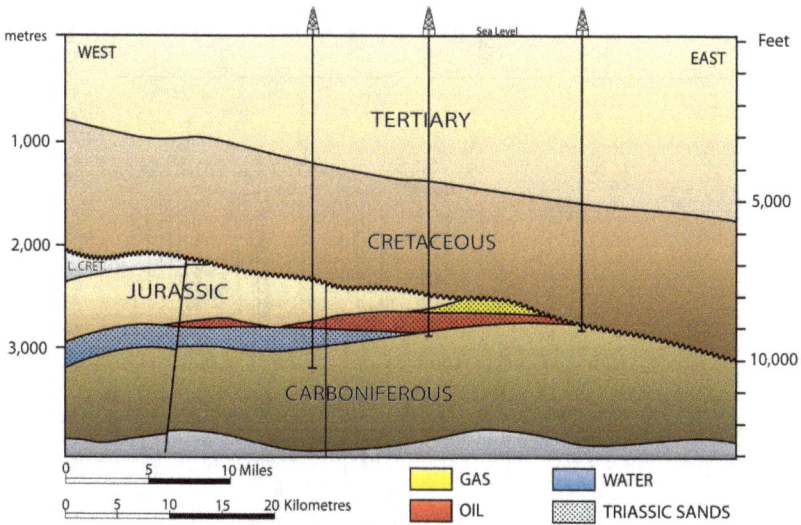

Figure 4.55. Cross-section of the Prudhoe Bay Oil Field, Alaska, an example of an unconformity- and fault-associated combination trap (modified and redrawn from Chapman, 1977).

the overlying strata were touched on in Chapter 1 and need not be repeated here. Salt diapirism has played a major role in creating oil and gas traps in several parts of the world, examples of which include the Gulf of Mexico coast of the US, Brazil, Gabon and the North Sea. In some cases the domes have risen from great depth, for example in the US Gulf of Mexico region, as shown in Fig. 4.56. Here, the salt layer is of Jurassic age and lies at a depth of more than 45,000 ft (14,000 m).

Figure 4.57 presents a diagrammatic illustration of the trapping opportunities created by a piercement dome. The traps develop over the top of the dome as well as along its flanks and include structural, stratigraphic and combination types.

- Upward arching of the strata immediately above piercement as well as non-piercement turtleback structures produces anticlinal traps as shown in Figs. 4.58 and 4.59.
- The tension resulting from upward arching causes normal faulting that, in turn, leads to the formation of fault-controlled traps. In

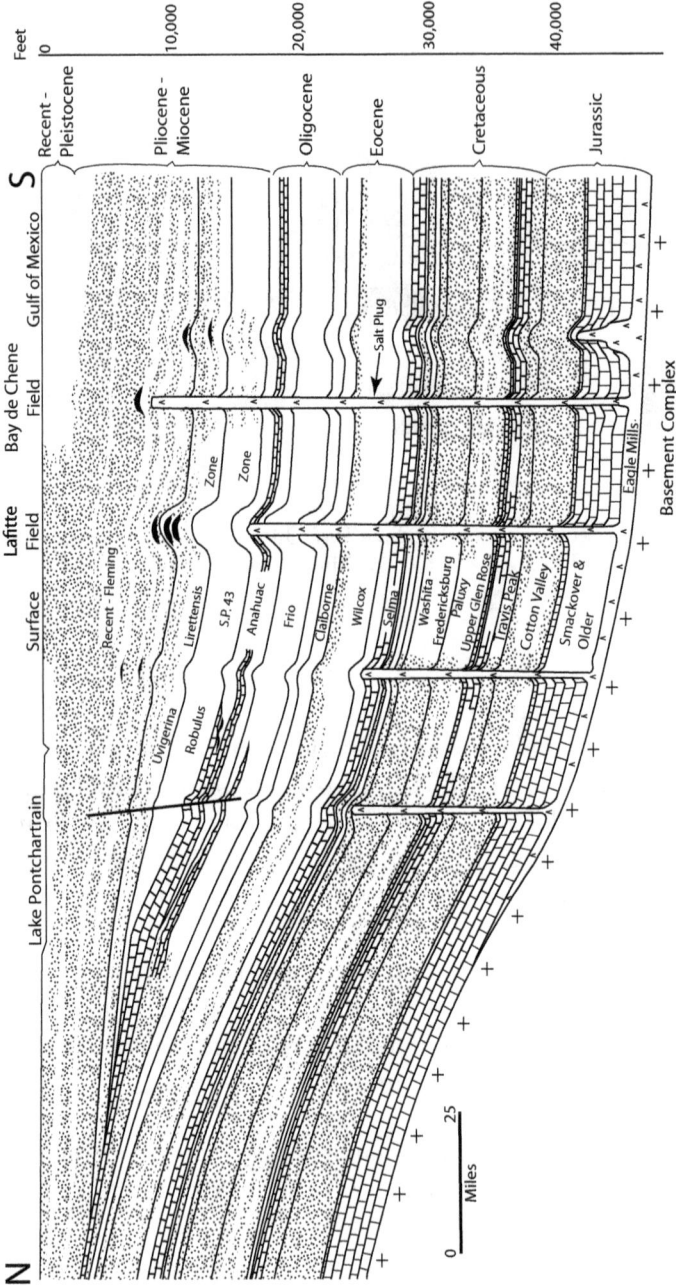

Figure 4.56. Diagrammatic section through the Mississippi coastal area, showing salt dome intrusions into the overlying section (modified and redrawn from Levorsen, 1967).

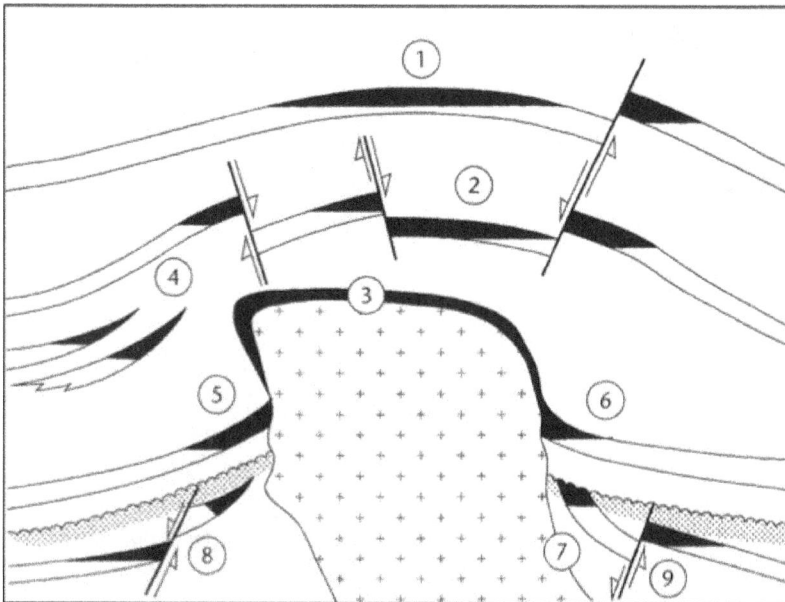

Figure 4.57. Diagrammatic illustration of trapping possibilities associated with a piercement salt dome (from Allen and Allen, 2005).

the seismic section shown in Fig. 4.60 fault development associated with the top of a piercement dome is clearly visible. The faults follow a complex and radial pattern, as depicted in Fig. 4.61.

- Fault-associated traps, related to an earlier phase of salt rise, can also be found along the flanks of the dome (Fig. 4.57).
- Pinch-out- and unconformity-associated traps resulting from uplift during an earlier phase of movement can also occur on the flanks (Figs. 4.57 and 4.58).
- Being impermeable, the salt is an effective barrier to lateral oil and gas movement and truncation of permeable beds by the salt mass can form traps, as illustrated in Figs. 4.57 and 4.58. These

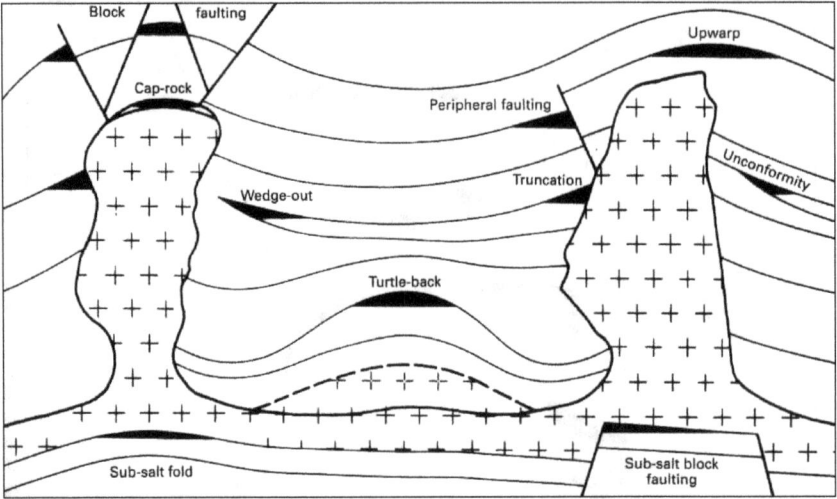

Figure 4.58. Diagrammatic illustration of the variety of trap possibilities with salt movement (after Stoneley, 1995).

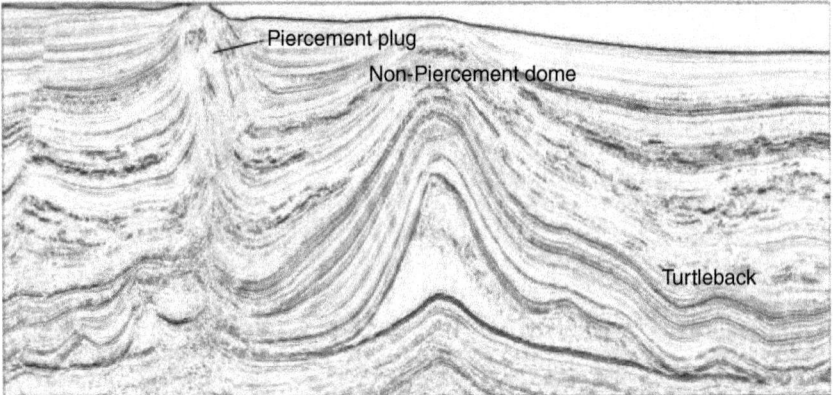

Figure 4.59. A seismic section, Kuanza Basin, offshore Angola, showing a piercement salt plug and a turtleback (http://www.cgg.com/data//1/rec_imgs/8409_Angola%20section%2033.jpg).

are called salt abutment traps and the Banff Field, central North Sea, is an example of this trap type (Fig. 4.62).

It should be noted that salt diapirism is involved only in the creation of traps and has no impact on the other petroleum system elements and processes.

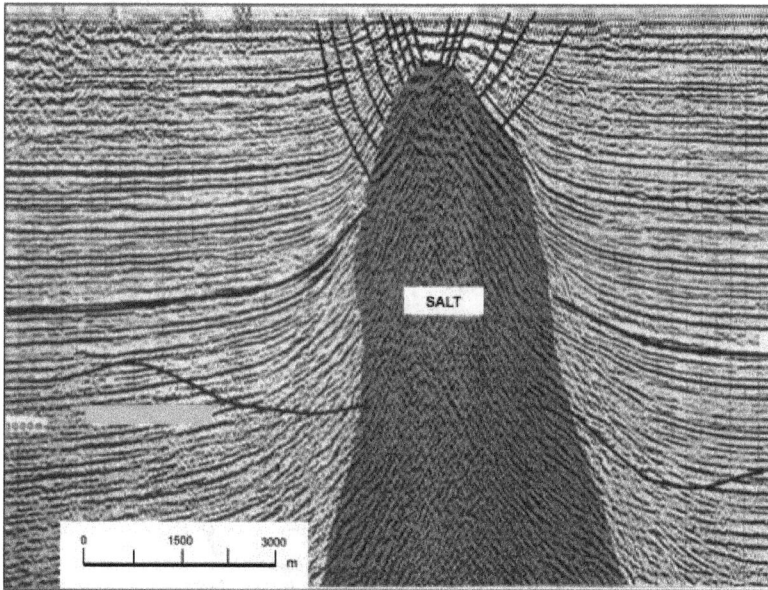

Figure 4.60. Seismic section showing normal faulting associated with the top of a piercement salt dome (From Lowell, 1985).

## 4.6. Seals

*Seals* or *cap rocks* are low permeability layers which lie on top of or adjacent to the reservoir, forming *top seals* or *lateral seals*, respectively. Their function is to prevent further movement and loss of the hydrocarbons from the trap. To be most effective, seals should be laterally extensive. Shales, mudstones, evaporites and sometimes unfractured limestones act as seals. Evaporites (salt and anhydrite) are the most effective cap rocks on account of their plastic properties, which cause the healing of any fractures resulting from folding or faulting, thereby preserving the integrity of the seal. Figure 4.63 provides an illustration of a top seal.

Shale seals are dominant in the northern North Sea fields, while evaporites cap the vast majority of the southern North Sea accumulations. An example of a northern North Sea oil accumulation capped by shale is the Ninian Oil Field, northern North Sea, a cross-section of which is shown in Fig. 4.64. Here, a Middle Jurassic sandstone reservoir is sealed by a thick succession of Upper Jurassic and Cretaceous shales.

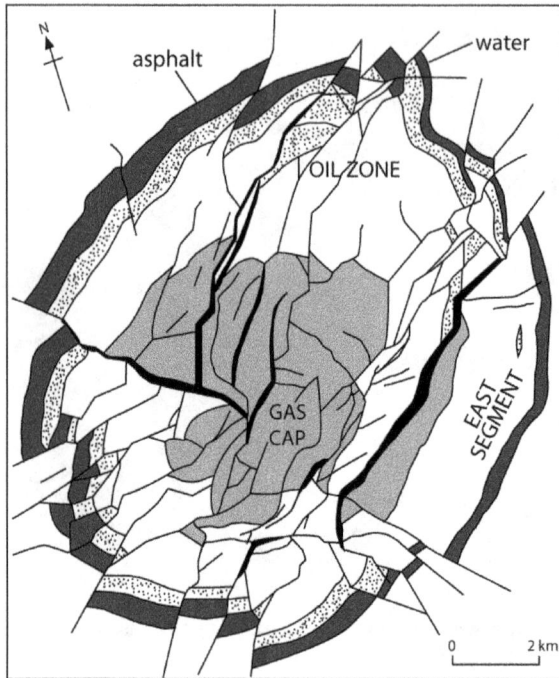

Figure 4.61. Outline of the Hawkins Field, Texas, a symmetrical dome over a deep salt uplift showing a complex radial fault pattern (modified and redrawn from North, 1985).

The Upper Permian Zechstein evaporites form a thick, laterally extensive layer in the southern North Sea and act an effective seal to the gas fields in this area. The Neptune Gas Field, southern North Sea is a good example of such an accumulation (Fig. 4.65; see also Fig. 4.45).

Many of the giant oil and gas fields in the Middle East and North Africa are also sealed by evaporites. Figure 4.66 presents a cross-section of the Hassi Messaoud Oil Field, Algeria, where sandstone reservoirs of Palaeozoic age are sealed by a Triassic evaporite layer (Fig. 4.66).

## 4.7.  The Petroleum Play

The term "play" is used extensively in the upstream sector (exploration and production activities) of the petroleum industry. It

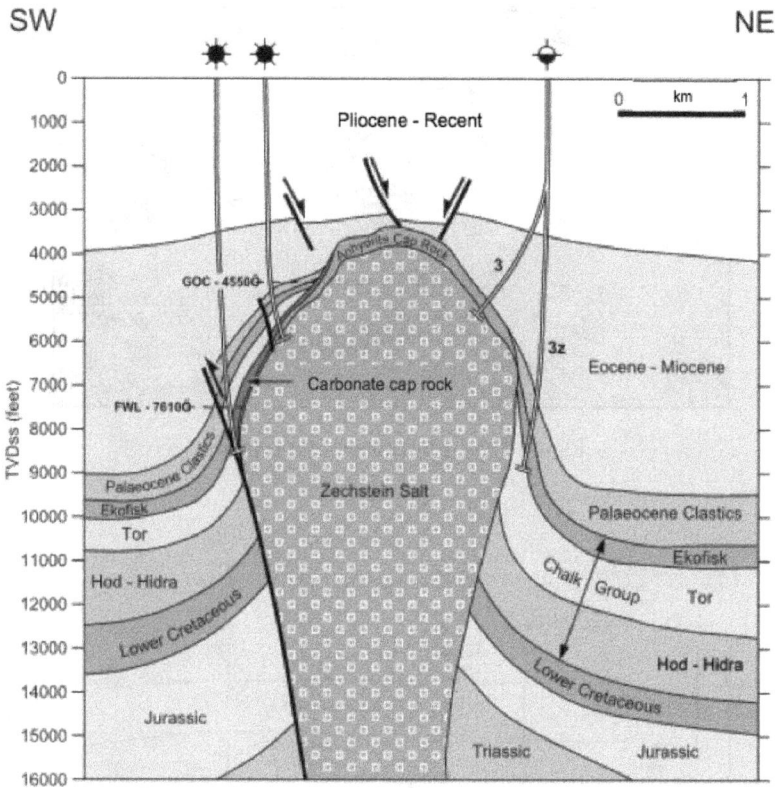

Figure 4.62. Cross-section through the Banff Field, central North Sea, showing fault-associated traps on the southwestern flank of a piercement salt dome (from Underwood, 2003).

is defined as a group of fields or drilling prospects in a given region, identified by common geological and engineering characters. The former include common source, reservoir, seal, trap, timing, migration and preservation features, while the latter embrace the fluid properties of the hydrocarbons and flow characteristics of the producing reservoirs. The concept of play provides a framework for understanding the distribution of the hydrocarbons in an area and is best demonstrated by reference to an example, such as the North Sea Basin.

Broadly, the North Sea can be divided into a southern gas-bearing basin and a northern province that contains both oil and gas fields.

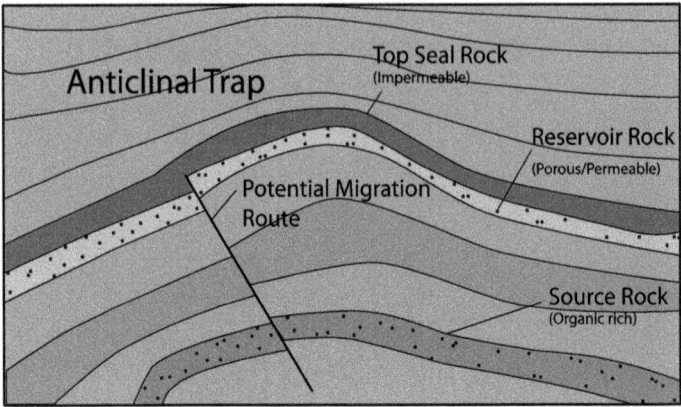

Figure 4.63. Illustration of a top seal.

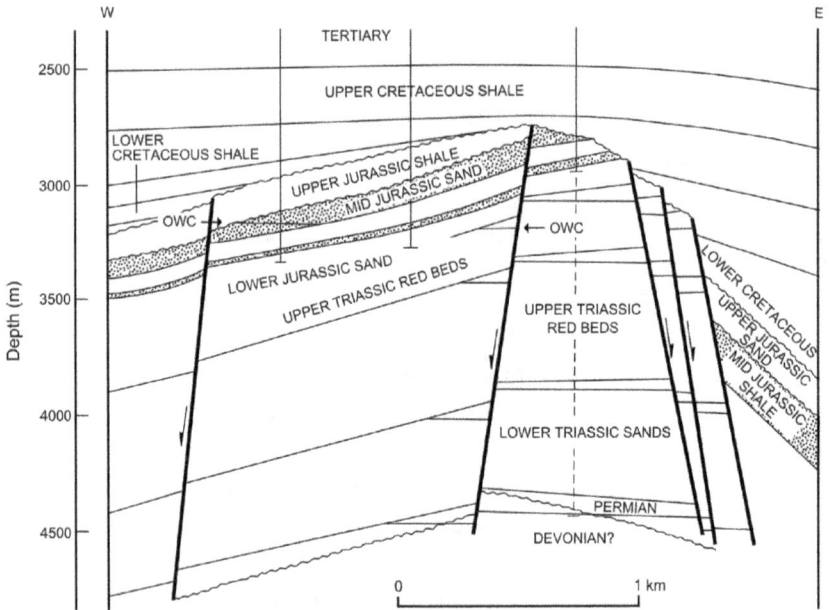

Figure 4.64. An east–west section across the Ninian Oil Field, northern North Sea, an example of an accumulation sealed by shale (modified and redrawn from North, 1985).

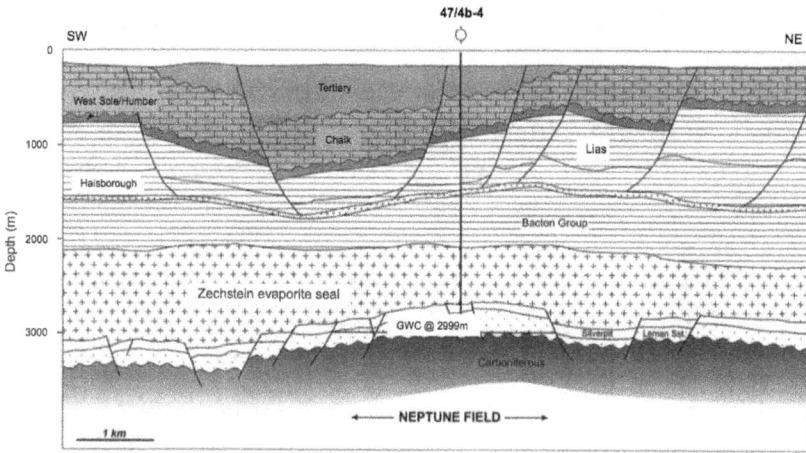

Figure 4.65. A northeast–southwest section across the Neptune Gas Field, southern North Sea, an example of an accumulation sealed by evaporites (after Smith and Starcher, 2003).

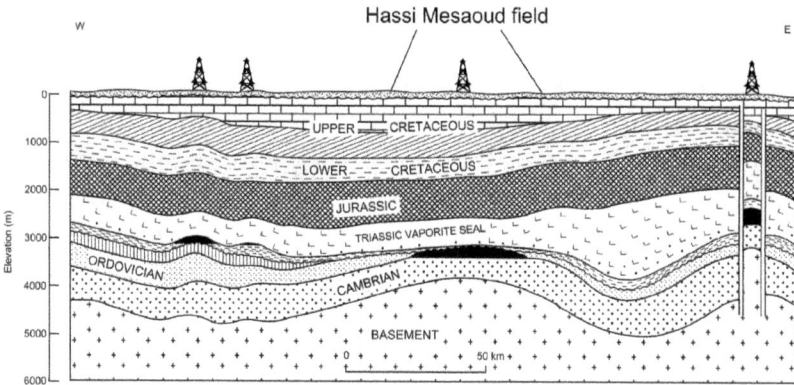

Figure 4.66. An east–west section across the Hassi Messaoud Oil Field, Algeria an example of an accumulation sealed by evaporites (modified and redrawn from North, 1985).

The southern North Sea is referred to as a "gas play", since the fields contain primarily dry gas; share the same reservoirs (with few exceptions), namely, the desert sands of Late Permian age (the Rotliegend sandstone); are charged by the same source, (i.e. the coals in the underlying Carboniferous successions); and have traps that are associated with horsts and tilted fault blocks and are sealed by the Late Permian Zechstein evaporites.

The northern North Sea plays are younger and more varied. There is a Jurassic "Brent play" where the reservoirs are formed by a series of deltaic sands of Middle Jurassic age and named after the discovery where they were first found (the Brent Field); they share the same source rock, namely, the overlying Upper Jurassic Kimmeridge clay in the UK sector and its equivalent, the Draupne shale in the Norwegian sector; the traps are associated with tilted fault blocks and are capped by interbedded shales.

Along the margins of the Viking and Central Grabens, an Upper Jurassic play becomes dominant. In this case the reservoirs are formed by sandstones that were deposited contemporaneously with the Kimmeridge clay. The reservoirs are charged by the Kimmeridge clay source rock; the traps are controlled by tilted fault blocks with a stratigraphic element in some cases and are sealed by the Kimmeridge clay and succeeding Cretaceous shales.

A very different play is developed in the Norwegian and Danish sectors, where chalk of Upper Cretaceous/Lower Tertiary (Palaeocene) age forms the reservoirs. Chalk is normally extremely fine-grained and has negligible permeability but it is able to function as a reservoir in this case on account of being fractured by salt diapirism. Rising from the underlying Permian Zechstein evaporites, the diapirs cause upward arching in the chalk layers, leading to the development of tension-related fractures. Also, the underlying Kimmeridge clay is over-pressured here and the high pressure is transmitted upward, helping to drive the oil through the chalk reservoirs into the wells, enabling commercially viable production rates to be achieved. The fields share the same Kimmeridge clay source rock; the traps are salt-dome-controlled and are capped by a thick layer of younger Tertiary shales.

Sandstones of Palaeocene and Eocene age, deposited as submarine fans in a deep-water environment, constitute a major play along the western side of the northern North Sea, mostly in the UK sector. The depositional model for this type of sand accumulation is illustrated in Fig. 4.67.

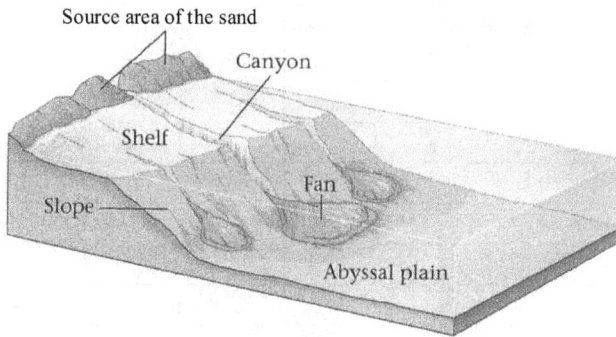

Figure 4.67. Depositional model for deep-water, fan-shaped sand accumulations (modified from Marshak, 2005).

Derived from the bordering highlands, the sand is deposited initially in the adjacent shelf area and reaches the deep marine environment, the abyssal plain, through episodic slides down canyons that cut into the shelf and slope. Such sediment bodies are fan-shaped, known as *turbidites* and result from deposition from *turbidity currents*, which are flows of water laden with mud, silt and sand. The reservoirs have been charged through faults and fractures by the Kimmeridge clay source rock; the traps are largely stratigraphic with an element of compaction drape in some cases and sealed by younger Tertiary shales.

## 4.8. The Petroleum Events Chart

Costs associated with exploration and drilling programmes began to escalate in the 1980s, making the oil industry more sensitive to risk. This led to demands by the industry for more robust information before committing to increasingly expensive projects and led to the development of basin and petroleum system modelling techniques. The techniques involve the integration of geological, geophysical, geochemical and geothermal data to assess whether favourable conditions for hydrocarbon generation, migration, accumulation and preservation exist in a given area. This can be achieved by constructing a 2D chart showing the distribution of the elements in time and dating the generation and migration processes. The result

| Time, millions of years ago (Ma) | | | | | | | Geologic time scale | | |
|---|---|---|---|---|---|---|---|---|---|
| 300 | | 200 | | 100 | | | | | |
| Paleozoic | | Mesozoic | | | Cenozoic | | | | |
| P | Per | Tr | J | K | Pg | Ng | Petroleum system events | | |
| | | | | | | | Source rock | Elements | |
| | | | | | | | Reservoir rock | | |
| | | | | | | | Seal rock | | |
| | | | | | | | Overburden rock | | |
| | | | | | | | Trap formation | Processes | |
| | | | | | | | Generation, migration, accumulation | | |
| | | | | | | | Preservation | | |
| | | | | | | | Critical moment | | |

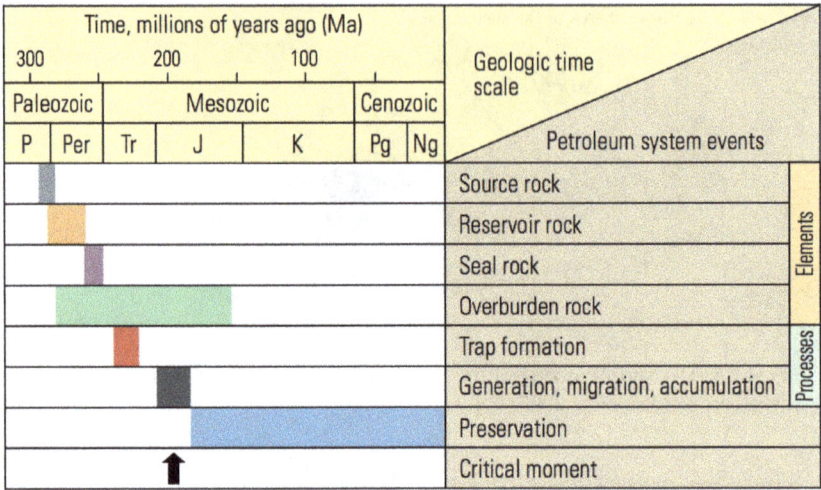

Figure 4.68. A generic petroleum system events chart. Each of the coloured horizontal bars represents the time span of an event. In this system all the essential elements are present, the processes have taken place and the timing is favourable; trap formation predates migration (copyright Schlumberger, after Al-Hajeri *et al.*, 2009, used with permission).

is referred to as a *petroleum events chart*, a generic example of which is presented in Fig. 4.68.

Each of the coloured horizontal bars on the chart represents the time span of an event. For this system, all the essential elements are present and their stratigraphic distribution is favourable; i.e. the source rock is older and therefore below the reservoir, the seal and overburden rocks are in the right place and the processes have occurred. The time of hydrocarbon generation, migration and accumulation, known as the *critical moment* and indicated by the black arrow, postdates trap formation, and the bar representing preservation shows that the accumulation has remained intact since the Middle Jurassic. The chart, therefore, represents a viable or functioning petroleum system in this case.

Figure 4.69 presents a petroleum events chart for the Prudhoe Bay Oil Field, Alaska. Favourable development of elements and timing of processes have resulted in the formation of a giant oil

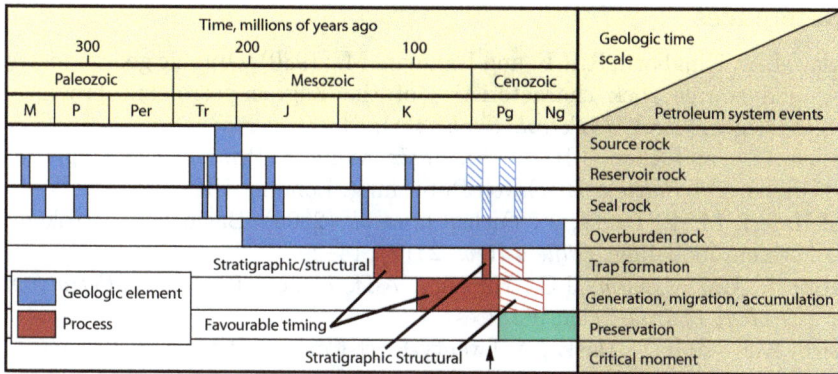

Figure 4.69. Petroleum events chart, Prudhoe Bay Oil Field, Alaska. Favourable development of elements and timing of processes have resulted in the formation of a giant oil field (copyright Schlumberger, redrawn after Al-Hajeri *et al.*, 2009, used with permission).

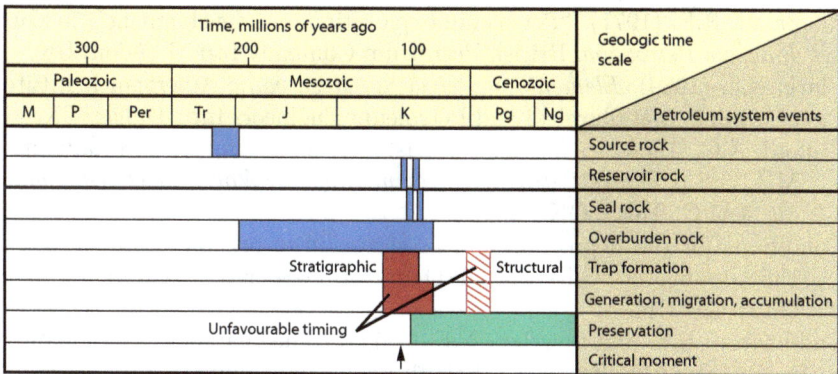

Figure 4.70. Petroleum events chart, Brooks Range Foothills, Alaska, showing less favourable timing of events and processes compared to Prudhoe Bay (copyright Schlumberger, redrawn after Al-Hajeri *et al.*, 2009, used with permission).

field. For comparison purposes, an events chart constructed for the Brooks Range Foothills, an area to the south of Prudhoe Bay, is shown in Fig. 4.70. Here, the timing of the events and processes were less favourable and this is reflected in the lower prospectivity of this region.

## References

Ala, M.A., Kinghorn, R.R.F. and Rahman, M. (1980). Organic geochemistry and source rock characteristics of the Zagros petroleum province, southwest Iran, *J. Petrol. Geol.*, **3**(1), 61–89.

Allen, P.A. and Allen, J.R. (2005). *Basin Analysis: Principles and Applications*, 2nd *Edition*. Blackwell Publishing, London, UK.

Al-Hajeri, M.M., Derks, J., Fuchs, J. *et al.* (2009). Basin and petroleum system modeling, *Oilfield Rev.*, **21**(2), 14–29.

Aqrawi, A.M.A., Goff, J.C., Horbury, A.D. *et al.* (2010). *The Petroleum Geology of Iraq*, Scientific Press, UK.

Bird, K.J. (2001). "Alaska, A Twenty-First-Century Petroleum Province", in Downey, M.W., Threet, J.C. and Morgan, W.A. (eds.), *Petroleum Provinces of the Twenty-First Century, AAPG Memoir 74*, AAPG, Tulsa, OK.

Brownfield, M.E. and Charpentier, R.R. (2006). Geology and total petroleum systems of the West-Central Coastal Province (7203), West Africa, *U.S. Geological Survey Bulletin 2207-B*. Available online at: http://www.usgs.gov/bul/2207/B/.

Chapman, R.E. (1977). "Petroleum exploration and development", in *Our Industry Petroleum*, British Petroleum Company Ltd., London, UK.

Clark, N.J. (1969). *Elements of Petroleum Reservoirs*, American Institute of Mining, Metallurgical and Petroleum Engineers Inc., Dallas, TX.

Clifford, A.C. (1986). "African oil — past, present and future", in Halbouty, M.T. (ed.), *Future Petroleum Provinces of the World, AAPG Memoir 40*, AAPG, Tulsa, OK.

Conditions for the deposition and preservation of organic matter (1996). United Kingdom Off shore Oil Operators Association (UKOOA), now Oil & Gas UK.

Cuddington, K.S. and Lowther, N.F. (1977). "The character of crude oil", in *Our Industry Petroleum*, British Petroleum Company Ltd.

Halbouty, M.T. (1972). "Rationale for deliberate pursuit of stratigraphic, unconformity and Paleogeormorphic traps", in King, R.E. (ed.), *Stratigraphic Oil and Gas fields — Classification, Exploration Methods and Case Histories, AAPG Memoir 16*, AAPG, Tulsa, OK.

Halbouty, M.T. and Barber, T.D. (1972). "Port acres and port arthur fields, Jefferson County, Texas: Stratigraphic and structural traps in a middle tertiary delta", in King, R.E. (ed.), *Stratigraphic Oil and Gas fields — Classification, Exploration Methods and Case Histories, AAPG Memoir 16*, AAPG, Tulsa, OK.

Henriksen, N. (2008). *Geological History of Greenland*, Geological Survey of Denmark and Greenland (GEUS), Copenhagen, Denmark.

Hobson, G.D. and Tiratsoo, E.N. (1985). *Introduction to Petroleum Geology*, 2$^{nd}$ *Edition*, Gulf Publishing Co., Houston, Texas.

Hunt, J.M. (1995). *Petroleum Geochemistry and Geology*, 2$^{nd}$ Edition, W.H. Freeman, New York.

Jassim, S.Z. and Al-Gailani, M.B. (2006). "Hydrocarbons", in Jassim, S.Z. and Goff, J.C. (eds.), *Geology of Iraq*, Dolin, Prague and Moravian Museum, Brno, Czech Republic.

Levorsen, A.I. (1967). *Geology of Petroleum*, 2$^{nd}$ Edition, W.H. Freeman, San Francisco.

Lindsay, R.F., Cantrell, D.L., Hughes, T.H. *et al.* (2006). "Ghawar Arab-D reservoir: Widespread porosity in shoaling-upward carbonate cycles, Saudi Arabia", in Harris, P.M. and Weber, L.J. (eds.), *Giant Hydrocarbon Reservoirs of the World: From Rock to Reservoir Characterisation and Modelling"*, *AAPG Memoir 88* Joint SEPM Publication.

Lowell, J.D. (1985). *Structural Styles in Petroleum Exploration*, Pennwell Corp., Tulsa, OK.

Marshak, S. (2001). *Earth: Portrait of a Planet*, 1$^{st}$ Edition, W.W. Norton, New York, NY.

Marshak, S. (2005). *Earth: Portrait of a Planet*, 2$^{nd}$ Edition, W.W. Norton, New York, NY.

McQuillan, H. (1973). Small scale fracture density in asmari formation of southwest Iran and its relation to bed thickness and structural setting, *AAPG Bull.*, **57**(12), 2367–2385.

North, F.K. (1985). *Petroleum Geology*, Allen and Unwin, London, UK.

Oil Service Company of Iran (1974). *1:100,000 Geological Map Series, Sheet 20842 (Fahliyan)*.

Osbon, R.A. Werngren, O.C., Kyei, A. *et al.* (2003). "The Gawain Field, Blocks 49/24, 49/29a, UK North Sea", in Gluyas, J.G. and HIchens, H.M. (eds.), *United Kingdom Oil and Gas Fields Commemorative Millennium Volume, Geological Society Memoir 20*, The Geological Society Publishing House, Bath, UK.

Petersen, H.I., Fyhn, M.B.W., Nielsen, L.H. *et al.* (2014). World class paleogene source rock from a cored Lacustrine Syn-rift succession, Bach Long Island Vi, Island, Song Hong Basin, Offshore Northern Vietnam, *J. Petrol. Geol.*, **37**(4), 373–389.

Rittenhouse, G. (1972). "Stratigraphic trap classification", in King, R.E. (ed.). *Stratigraphic Oil and Gas fields — Classification, Exploration Methods and Case Histories, AAPG Memoir 16*, AAPG, Tulsa, OK.

Selley, R.C. (1997). *Elements of Petroleum Geology*, 2$^{nd}$ Edition. Academic Press, Waltham, MA.

Sellwood, B.W. (2007). *Carbonates*, Department of Earth Science and Engineering, Imperial College London.

Smith, B. and Stracher, V. (2003). "The Mercury and Neptune Fields, Blocks 47/9b, 47/4b 47/5a, 42/29, UK North Sea", in Gluyas, J.G. and HIchens, H.M. (eds.), *United Kingdom Oil and Gas Fields Commemorative Millennium Volume, Geological Society Memoir 20*, The Geological Society Publishing House, Bath, UK.

Southeast Zagros Basin, Iran, satellite image. Available online at: http://en.wikipedia.org/wiki/Persian_Gulf.

Stoneley, R. (1995). *An Introduction to Petroleum Geology for Non-Geologists*, Oxford University Press, Oxford, UK.

Underhill, J.R. (2003). "The tectonic and stratigraphic framework of the United Kingdom's oil and gas fields", in Gluyas, J.G. and Hichens, H.M. (eds.), *United Kingdom Oil and Gas Fields Commemorative Millennium Volume, Geological Society Memoir 20*, The Geological Society Publishing House, Bath, UK.

# Chapter 5

# Exploring for Oil and Gas

## 5.1. Introduction

The first step in the assessment of the hydrocarbon potential of an area is the collection of data. Broadly, the information required falls into three categories: geological, remote sensing and geophysical. These are described briefly below.

## 5.2. Geological Sources

Geological sources include data in the published literature and maps and reports produced by national geological survey organisations. There is also usually some unpublished material in the form of commercial reports that is available for purchase. An example of a geological map is shown in Fig. 5.1. The colours represent different rock units or formations. A geological map is normally accompanied by a stratigraphic column (not present in this example) that identifies the rock units in terms of their local names, age, thickness and lithology, and a cross-section that illustrates the structures present (e.g. folds, faults, thrusts, etc.).

Where good exposures are present, as exemplified in the photograph in Fig. 5.2, studies of rock outcrops provide useful information on the structural configuration of the area as well as the occurrence and quality of the petroleum system elements: source, reservoir, trap and seal.

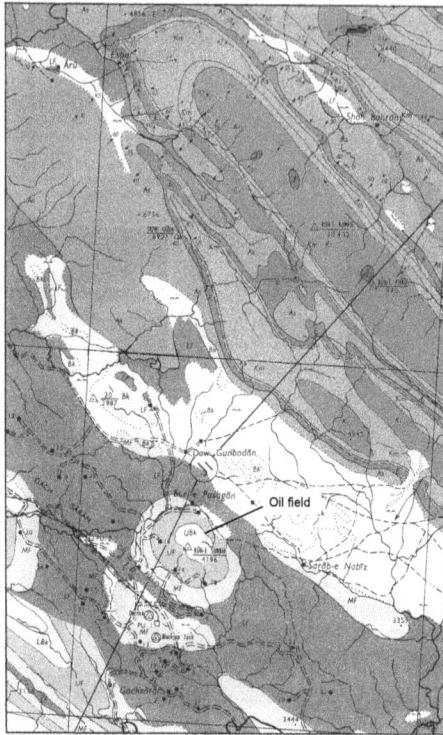

Figure 5.1. An example of a geological map and accompanying cross section (after Fox, 1970).

## 5.3.  Remote Sensing

Aerial photography and satellite imagery are the two most commonly forms of remote sensing sources of information.

Aerial photographs are taken looking vertically downwards from aircraft flying along a predetermined grid. They offer a rapid and inexpensive means of gathering information, can be taken at various scales and provide sufficient overlap between successive photos so that they can be viewed with a stereoscope, which provides a 3D picture of the features on the ground. Aerial photos yield a great deal of information about the geology and can be used in mapping surface structures; however, their use is limited to regions of good bedrock

Figure 5.2. An area of good exposure: outcrops showing steeply dipping strata (photo by J. Cosgrove).

(after Lowell, 1985)  (from Marshak, 2005)

Figure 5.3. Aerial photographs showing two large, elongated anticlines.

exposure. Figure 5.3 shows examples of aerial photos in which large, elongated anticlines are clearly evident.

Satellite images for the entire world are available at various scales. They are inexpensive to acquire and are useful in delineating large-scale features such as an entire country, the areal extent of onshore

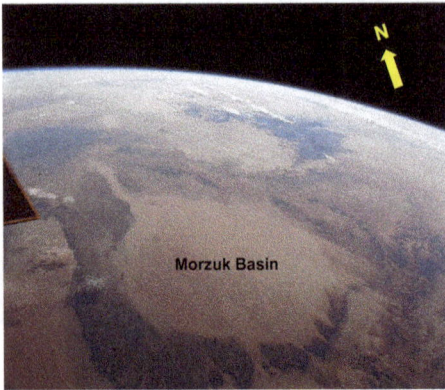

(a) Morzuk Basin, Libya. Surrounding dark patches are outcrops of Palaeozoic and Mesozioc rocks (earthobservatory.nasa.gov).

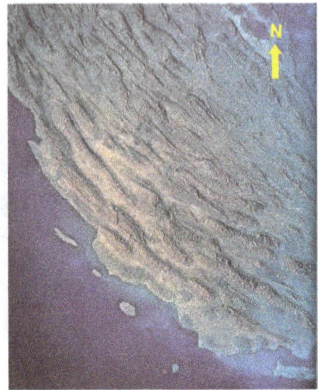

(b) Southeast sector, Zagros Basin, SW Iran. Note the NW–SE-trending, elongated folds (formerly at http://en.wikipedia.org/wiki/Persian_Gulf).

(c) Southeast sector, Zagros Basin, SW Iran. Dark patches are piercement salt (http://www.sciencephoto.com/media/173889/view).

Figure 5.4. Examples of satellite images of North Africa and the southwestern sector of the Zagros Basin in southern Iran.

sedimentary basins or regional structure, and in revealing major fold and fault patterns. Examples of satellite images covering North Africa and the southeastern sector of the Zagros Basin in southern Iran are presented in Fig. 5.4. The outline of the Morzuk Basin, Libya, is readily apparent in Fig. 5.4(a) and the regional northwest–southeast trend of the elongated anticlines of the Zagros fold belt can be seen clearly in Figs. 5.4(b) and 5.4(c).

## 5.4.  Geophysical Techniques

Geophysical techniques include gravity, magnetic and seismic surveys and offer the only means of obtaining information in the offshore or areas with no bedrock exposure. They respond to certain rock properties and the interpretation of the results provides information on the sub-surface geology.

### 5.4.1.  *Gravity surveying*

*Gravity surveying* is the oldest geophysical method. The tool, known as a *gravimeter*, was developed in the latter half of the nineteenth century and, with modifications, has been in use ever since. In gravity surveying, variations in $g$, the acceleration due to gravity, are measured and recorded in milligals. Nowadays, gravity surveying tends to be airborne, with the instrument being mounted on an aircraft, as shown in Fig. 5.5.

Figure 5.5.  Airborne magnetic survey aircraft (after Henriksen, 2008).

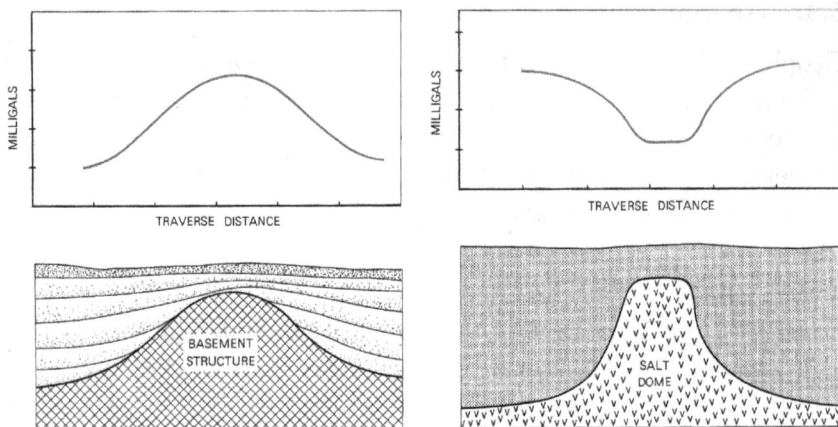

Figure 5.6. Variations in gravity due to geological factors. Positive anomalies indicate dense rocks close to the surface while negative anomalies represent low density rocks near the surface (after Piggott, 1977).

The variations in gravity are caused by changes in the density of the rocks in the sub-surface; heavy rocks will cause the measured value of $g$ to be higher than that associated with the presence of less dense ones. Deviations of the measured values from the regional average are referred to as *anomalies*. A positive anomaly indicates the presence close to the surface of dense material, such as igneous and metamorphic rocks, and may be the result of uplifts in the basement below the sedimentary cover. A negative anomaly suggests the presence of relatively low-density rocks, such as salt, near the surface. Figure 5.6 provides a diagrammatic illustration of positive and negative anomalies and their interpretation. Salt diapirs are thus readily identifiable by their negative anomaly signatures and, until the introduction of the seismic method after the First World War, gravity surveys were used to detect salt-dome-associated traps in Europe and the US Gulf Coast area.

Positive gravity anomalies are produced also by surface features such as mountain ranges. A gravity map of the southwestern US is shown in Fig. 5.7, showing particularly strong positive anomalies associated with the Sierra Nevada and Coast Range mountains in California. Negative anomalies are seen just offshore and on land where they correspond to onshore basins in California.

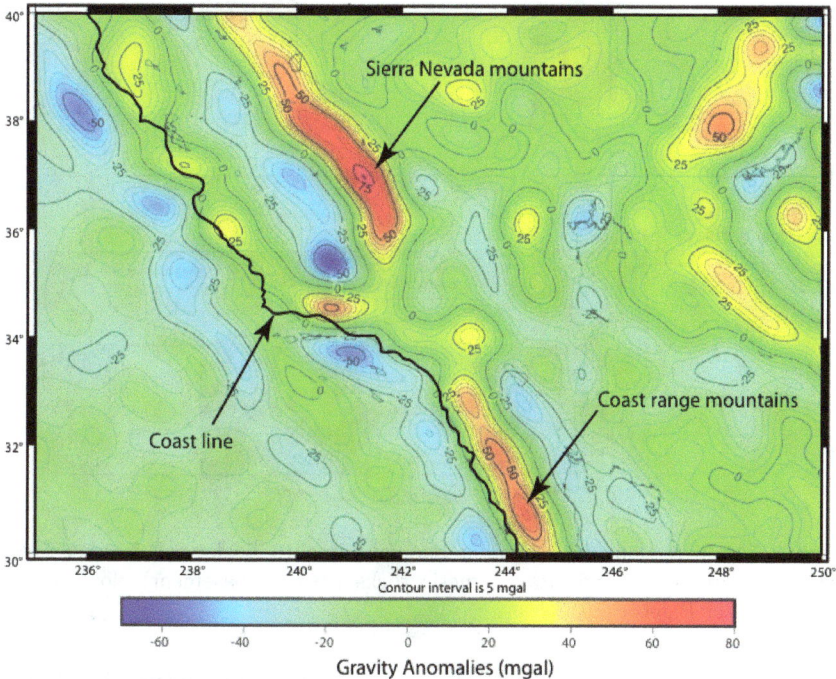

Figure 5.7.  Gravity anomaly map of southwest USA (http//www.macpurity.net/ geodsci/SWUS.htm).

It should be noted that gravity surveying is used by the oil industry for regional, reconnaissance purposes; it is particularly useful in the preliminary stages of an exploration programme to identify areas of interest, which are then covered by seismic surveys.

### 5.4.2.  *Magnetic surveying*

Certain rocks have magnetic properties and their presence has a local effect on the intensity and direction of the Earth's magnetic field. Like the variations in gravity, these magnetic effects can be measured and recorded by an instrument called a *magnetometer*. *Magnetic surveys* are airborne, with the device carried on an aircraft that flies along a predetermined grid. The data obtained is used to map the distribution of magnetic anomalies over the area of interest.

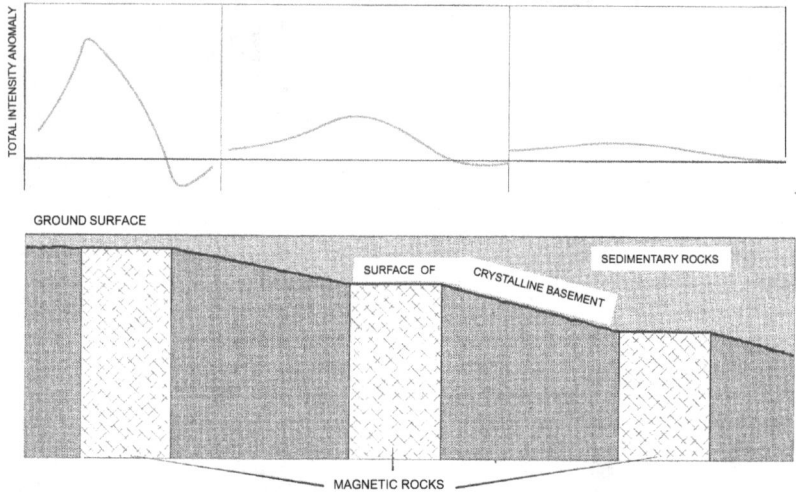

Figure 5.8. Variations in magnetic anomaly due to geological factors. High amplitude anomalies indicate magnetic rocks (usually basement) close to the surface (after Piggott, 1977).

Since sedimentary rocks are generally non-magnetic, the depth below surface of magnetic rocks, such as igneous and metamorphic complexes forming crystalline basement, controls the sharpness of the anomalies, as illustrated in Fig. 5.8. As an example, a magnetic anomaly map of the northern Gulf of California is presented in Fig. 5.9. The Altar Basin, bounded by two major northwest–southeast trending faults, stands out as a magnetic anomaly low flanked by magnetic anomaly highs that reflect a relatively shallow crystalline basement to the northeast and southwest. Magnetic anomaly maps, therefore, are useful regional indicators of the depth to basement, which in turn, allows the total thickness of the sedimentary section in a basin to be estimated. Major structural features can also be identified from these maps.

### 5.4.3.  *Seismic surveys*

Seismic surveying is the most important geophysical method and is used at various stages of both onshore and offshore exploration programmes.

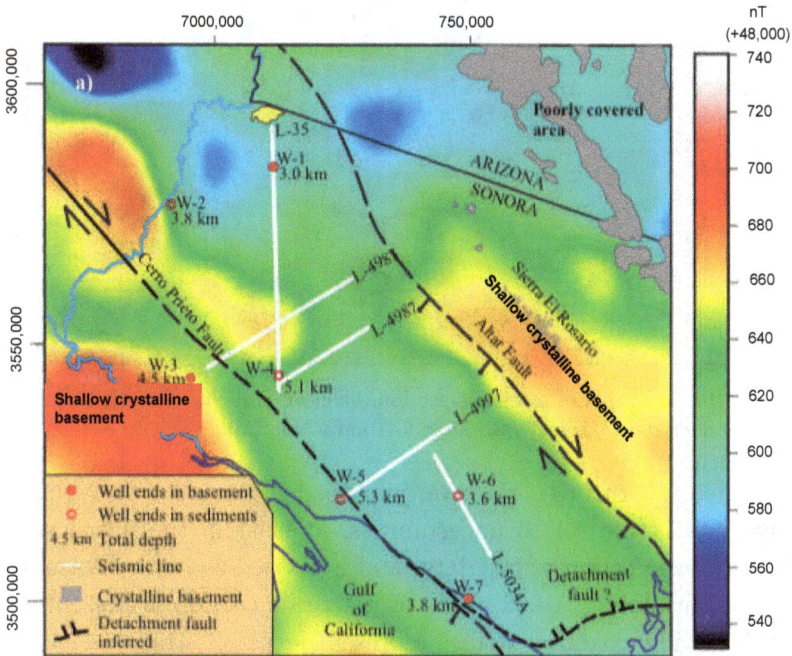

Figure 5.9. A magnetic anomaly map of the northern Gulf of California area. The Altar Basin stands out as a magnetic anomaly low flanked by magnetic anomaly highs that reflect a relatively shallow crystalline basement to the northeast and southwest (after Pacheco *et al.*, 2006).

Seismic data acquisition involves directing acoustic energy at the sub-surface rocks and measuring the arrival times of the waves that return to the surface. The time measured is that taken by the energy to travel from the source to the reflecting layer and back to the surface. It therefore represents a two-way travel time, measured in seconds, and abbreviated to TWT. This information is then processed to produce the familiar seismic sections, which provide a 2D picture of the sub-surface rock layers. It should be emphasised, however, that these are not geological cross-sections since the "depth" is plotted in TWT rather than the actual distance from the surface. Knowledge of the velocity of the seismic waves through the overlying layers is necessary to convert this time record to actual depths. The process is known as *depth conversion* but since

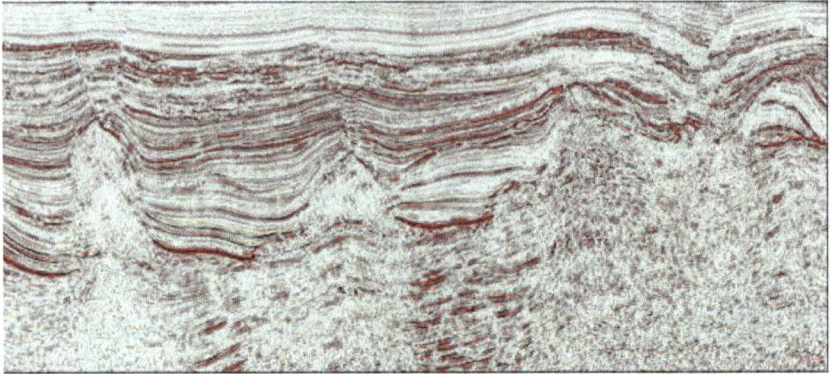

Figure 5.10. Example of a seismic section, offshore Gabon, west Africa (http://www.cgg.com/data//1/rec_imgs/8426_GAbon%20data.jpg).

wave velocity through the overlying beds is usually variable, the conversion is subject to uncertainties, at least in the initial stages of the interpretation of the data, and is revised as more information becomes available. An example of a seismic section is presented in Fig. 5.10. The coloured stripes represent layers of strata.

Onshore, the equipment is mounted on trucks, examples of which are shown in Fig. 5.11. Seismic energy is generated by exploding a charge of dynamite in a shallow-bore hole, dropping a heavy weight onto the ground or using the *vibroseis* technique, which involves shaking the ground by pressing a vibrating steel plate on the surface. Figure 5.12 presents a photograph of a vibroseis truck.

The waves propagate downwards and are reflected back at sub-surface rock boundaries at points where the layer properties change. These reflections from the sub-surface are detected and recorded by receivers at the surface, as illustrated in Fig. 5.13. On land, the receivers are called *geophones* (Fig. 5.14), and are arranged in lines or in a grid.

Offshore, the receivers are called *hydrophones* and are towed behind the survey vessel, as shown in Fig. 5.15. Due to its detrimental effect on marine life, dynamite is rarely used as a seismic energy source and has been replaced by the air gun, which generates seismic energy through small explosions by discharging compressed air from a chamber. Marine seismic data acquisition is illustrated in Fig. 5.16.

Figure 5.11. Seismic survey trucks (www.edgo.com).

Figure 5.12. A vibroseis truck (www.pgecurrents.com).

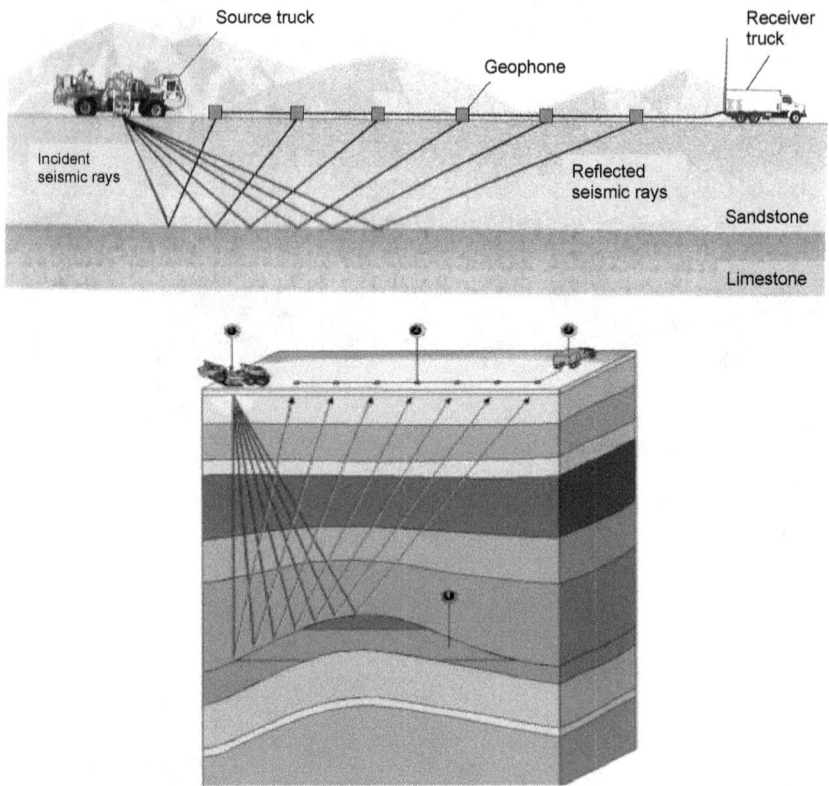

Figure 5.13. Principle of onshore seismic data acquisition (after Marshak, 2005).

It should be mentioned that offshore seismic surveying is more rapid and less expensive than land operations since it is a continuous process and no time is spent on moving from station to station and redeploying geophones. Deploying the geophones in built-up areas is particularly problematic and time-consuming in land surveys.

5.4.3.1. *Propagation of waves at an interface*

At an interface, the energy of the incoming waves is partitioned, as shown in Fig. 5.17: some continues along the interface and is referred to as the *refracted wave*; some is reflected and is called the *reflected wave*; and the remainder is transmitted across the interface and is designated the *transmitted wave*. It should be noted that the energy

Figure 5.14. A geophone used in land seismic surveys (courtesy Sercel).

Figure 5.15. A marine seismic survey operation (www.iagc.org).

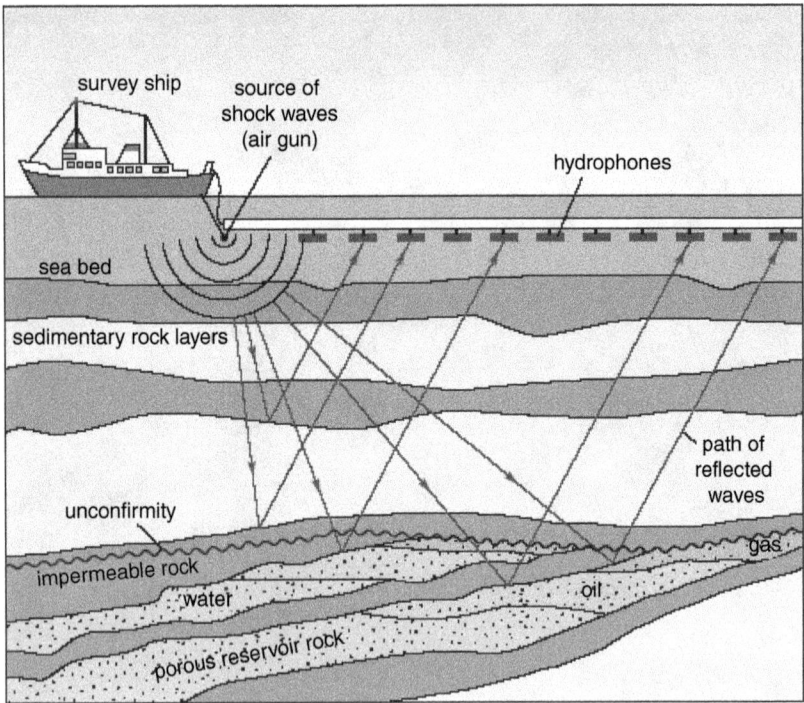

Figure 5.16. Marine seismic data acquisition (www.rigzone.com).

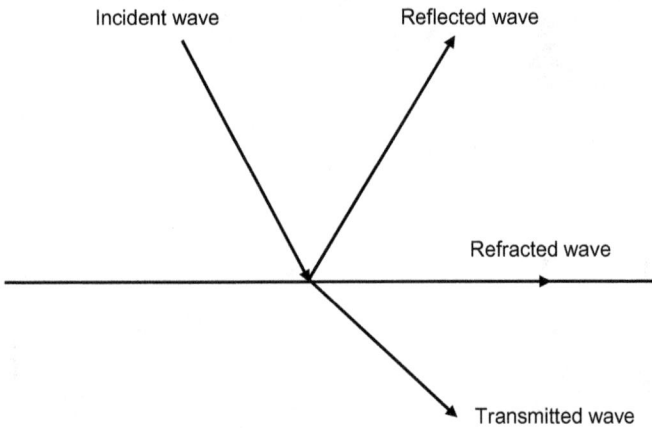

Figure 5.17. Propagation of waves at an interface.

**Layer 1**
Compressional wave *velocity* = $V_{p1}$
Density = $\rho_1$

_____   Interface between layers and 1 and 2

**Layer 2**
Compressional wave velocity = $V_{p2}$
Density = $\rho_2$

At the interface    $R = \dfrac{V_{p2}\rho_2 - V_{p1}\rho_1}{V_{p2}\rho_2 + V_{p1}\rho_1}$

Since $V_p\rho = I_a$,

$$R = \frac{I_{a2} - I_{a1}}{I_{a2} + I_{a1}}$$

Figure 5.18. Calculation of the coefficient of reflection at an interface between two layers.

of the refracted, reflected and transmitted waves is a fraction of that of the incident wave. The reflective properties of an interface depend on a factor called the *acoustic impedance* of the two media separated by the interface. This is abbreviated to $I_a$ and defined as

$$I_a = V_p \times \rho, \tag{5.1}$$

where $V_p$ is the compressional wave velocity and $\rho$ is the density.

The amount of energy reflected at an interface is called the *coefficient of reflection*, abbreviated to $R$. Considering the simple two-layer case depicted in Fig. 5.18, the coefficient of reflection at the interface separating them can be calculated as shown in the figure.

The greater the contrast between $I_{a1}$ and $I_{a2}$, the stronger the reflections from the interface, resulting in a better definition of the layers. Differentiation between layers with similar $I_a$ characteristics becomes uncertain.

5.4.3.2.   *Types of seismic survey*

Seismic surveys fall into the 2D, 3D, and 4D categories.

- *2D surveys* use 2–3 km grid spacing and provide regional coverage.
- *3D surveys* use 20–30 m grid spacing. Thousands of geophones/hydrophones are deployed and provide detailed information to define drilling objectives. They are expensive to run and the acquisition of 3D seismic data is limited to areas found to be interesting from the interpretation of regional or reconnaissance 2D coverage. Figure 5.19 illustrates the difference in detail between the results of 2D and 3D surveys.
- *4D surveys.* The "4$^{\text{th}}$ dimension" in 4D surveys is time. A 4D survey means that at least two 3D surveys have been run at different times over the same area, usually a field. The reflection characteristics of a reservoir change through time due to changes in fluid saturations. 4D surveys can therefore be used to monitor fluid movements in a reservoir.

5.4.3.3.   *Seismic data processing*

This is a specialised subject and can only be touched on in the context of this book. The original traces recorded by the geophones are not really usable as they are and need to be treated in various ways to produce seismic sections. The aim of seismic data processing is to emphasise primary reflections and to reduce spurious signals, which are referred to generally as *noise*. As mentioned earlier, the energy of waves returning to receivers is small compared to the source energy. This is due to energy loss during the downward as well as the upward propagation of the seismic waves; the weak signals are reinforced by adding together signals from different receivers to upgrade the quality of the data. This operation is known as *stacking*.

Figure 5.20 illustrates some of the terminology associated with processing. It shows a single source, numbered 21 on the diagram, shooting into multiple receivers and highlights a particular problem: the fact that reflections take longer to reach a distant geophone than one nearer the source. Consequently, a horizontal reflector will appear

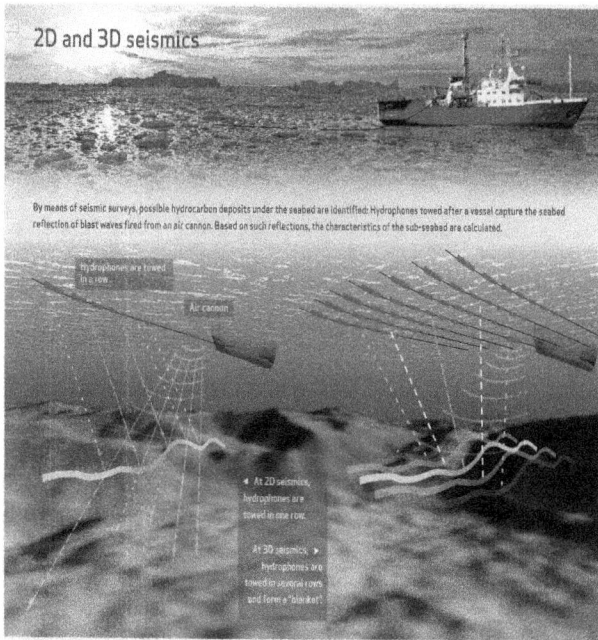

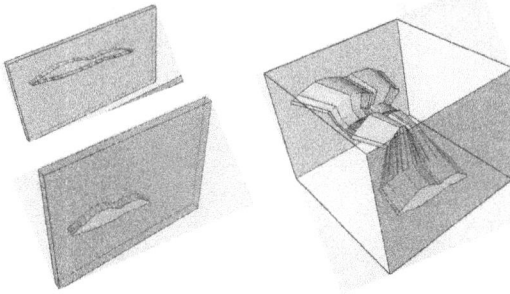

Figure 5.19. Comparison between the results of 2D and 3D seismic data (Bureau of Minerals and Petroleum, Greenland, 2009).

as a curved surface, becoming progressively deeper towards the more distant geophones, as shown in Fig. 5.21. This effect is known as *normal move out*, abbreviated to NMO, and must be corrected for. The case of a dipping reflector is illustrated in Fig. 5.22. Here, the ray paths are not symmetrical with respect to the source and the associated NMO correction is shown in Fig. 5.23.

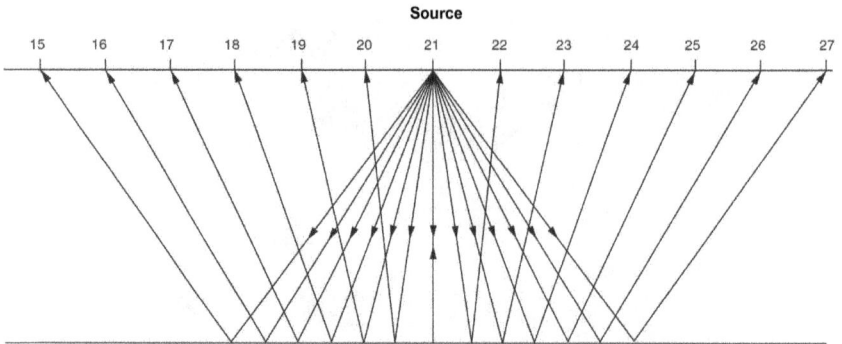

- The ray paths from a single source point shooting into multiple receiver locations is known as a *gather.*
- The distance between source and receiver is referred to as *offset.*
- Increasing travel time causes the reflector surface to be plotted as a hyperbola (see Fig. 5.21).
- Relocating the reflections to positions where they should be in a seismic section is called *migration.*
- Producing a *depth* section is the end goal of migration.

Figure 5.20. Ray paths from a single source to multiple receivers and the associated terminology.

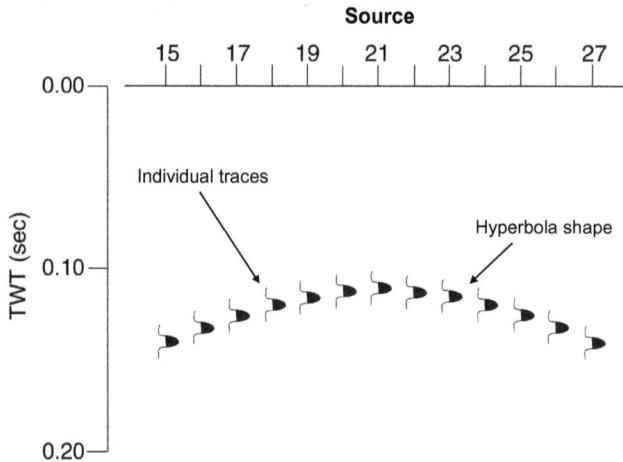

Figure 5.21. NMO correction for a horizontal surface.

## 5.4.3.4.  *Seismic interpretation*

The first step in the interpretation of seismic data is to pick prominent reflections, follow them along the section and identify structural features such as folds and faults. Nowadays the interpretation is

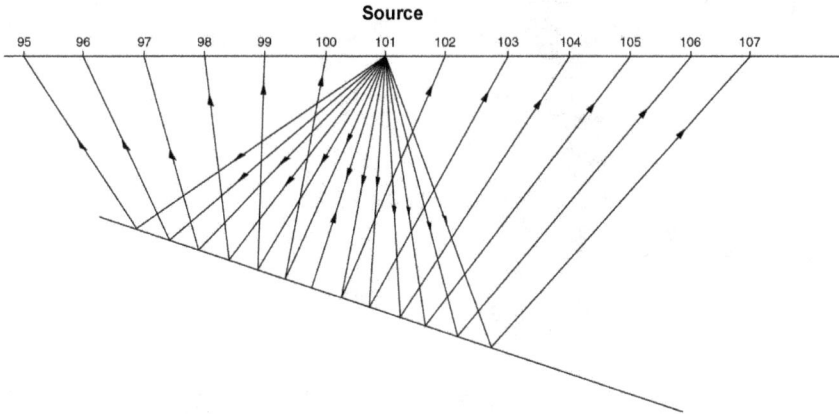

Figure 5.22. Ray paths from a single source to multiple receivers in the case of a dipping reflector.

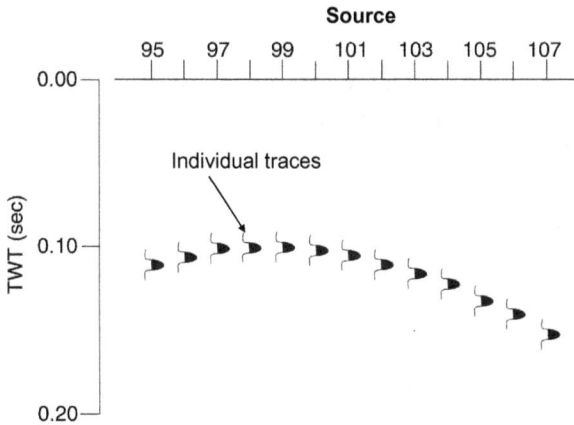

Figure 5.23. NMO correction for a dipping reflector.

computer-based, undertaken on dual screen workstations that allow the integration of seismic and well data to produce a wealth of information on many features of the sub-surface rocks. Commonly, tops of reservoirs produce strong reflections, making it possible to identify structural traps, anticlines in particular. Strong reflections also are often associated with unconformities.

Figure 5.24 presents a seismic section of the eastern flank of the giant Statfjord Field, in the Norwegian sector of the North Sea.

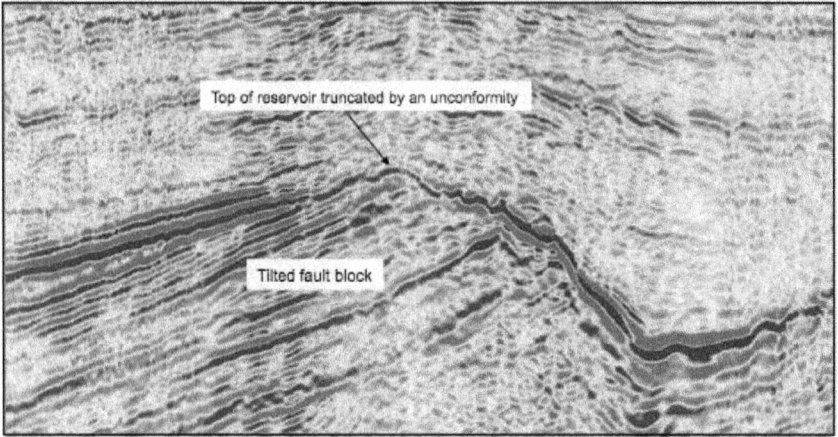

Figure 5.24. Seismic section, eastern flank, Statfjord Field, Norwegian sector, North Sea (from Bjorlykke, 2010).

The tilted-fault-block-controlled trap and the top of the reservoir truncated by an unconformity are readily identifiable.

Salt domes produce an "opaque" seismic response. This is due to the salt having lost its original bedding as the result of flowage. Two cross-seismic lines from the offshore Persian Gulf, an area of strong salt-dome activity, are presented in Fig. 5.25. The salt diapirs stand out as structureless masses, while the surrounding sedimentary successions are characterised by strong, laterally continuous reflections. Of particular interest is the undrilled salt-dome-controlled anticline with four-way dip closure. In such a richly petroliferous province, this would be a highly attractive drilling target.

Geological cross-sections may be derived from depth-converted seismic sections. This is a very useful application of seismic data and two such examples are presented in Figs. 5.26 and 5.27.

There have been significant advances in recent decades in 3D seismic data interpretation techniques and this remains an active area of research and development. These techniques have opened up new horizons in defining not only the structural configurations of the subsurface layers with greater accuracy, but also in providing clues to their geological characteristics. The latter studies are called *seismic stratigraphy*, the principles of which are discussed briefly below.

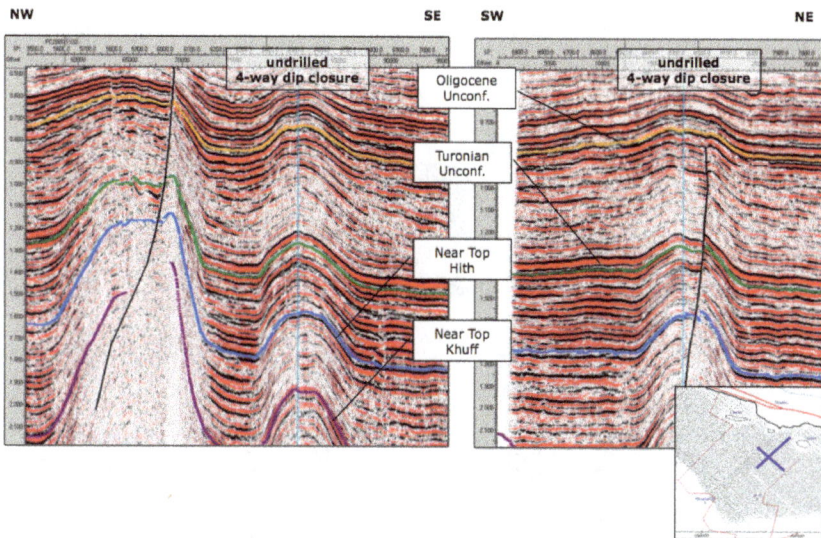

Figure 5.25. Two 2D seismic cross-lines, offshore Persian Gulf, showing an undrilled, salt-dome-controlled anticline with four-way dip closure. Note the poor seismic response from the diapirs (Global Geo Services AS, 2000).

A useful development in interpretation techniques makes it possible under certain conditions to detect hydrocarbons directly in the sub-surface. Contrasts in the seismic wave propagation properties of liquid- and gas-filled pore spaces in sandstone reservoirs sometimes cause reflections from gas–liquid contacts. On seismic sections this gives rise to horizontal anomalies known as *flat spots* and referred to as *direct hydrocarbon indicators* (DHIs). Figure 5.28 shows a seismic section across the Troll Gas Field, offshore Norway, on which a 7-km-long flat spot is evident at 1,700 ms. There is also extensive faulting in the reservoir.

There is insufficient contrast between the seismic wave propagation characteristics of oil- and water-filled pore spaces to allow flat spots to be developed at oil–water contacts.

### 5.4.3.5. *Seismic stratigraphy*

As noted in Chapter 1, in the stratigraphic sense, a sequence is a package of related strata bounded by unconformities. In seismic

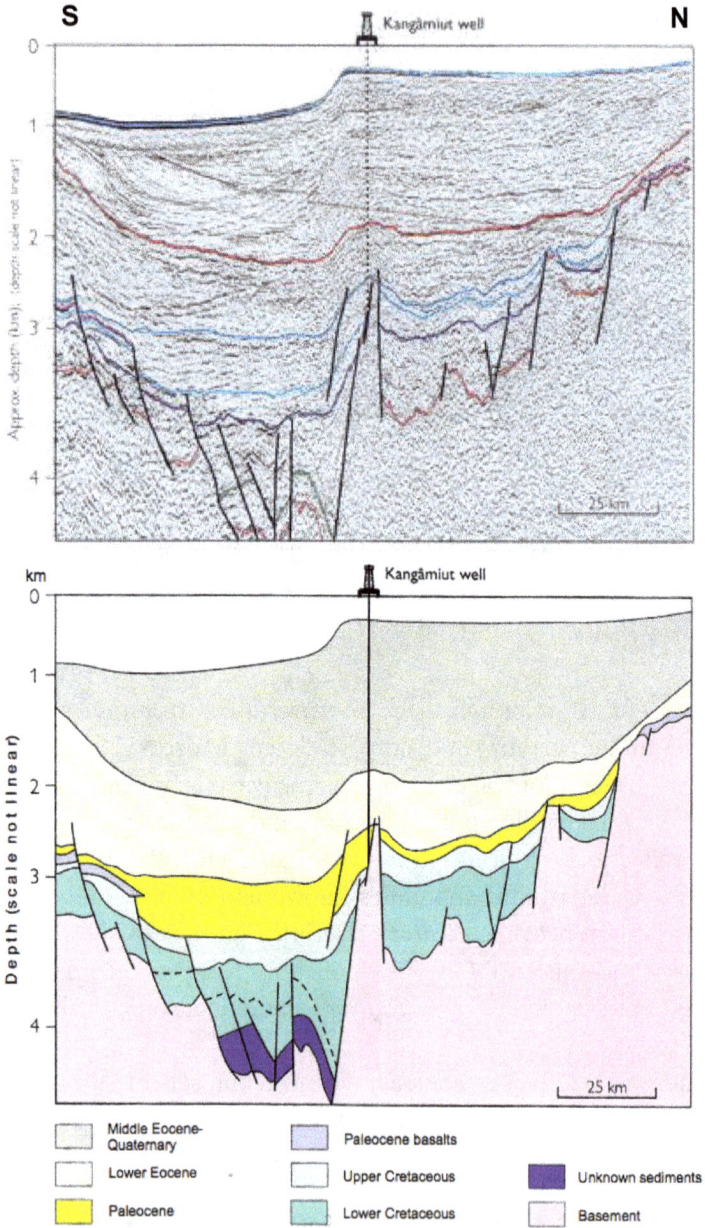

Figure 5.26. Example of the conversion of an interpreted seismic section into a geological cross section, offshore west Greenland (after Henriksen, 2008).

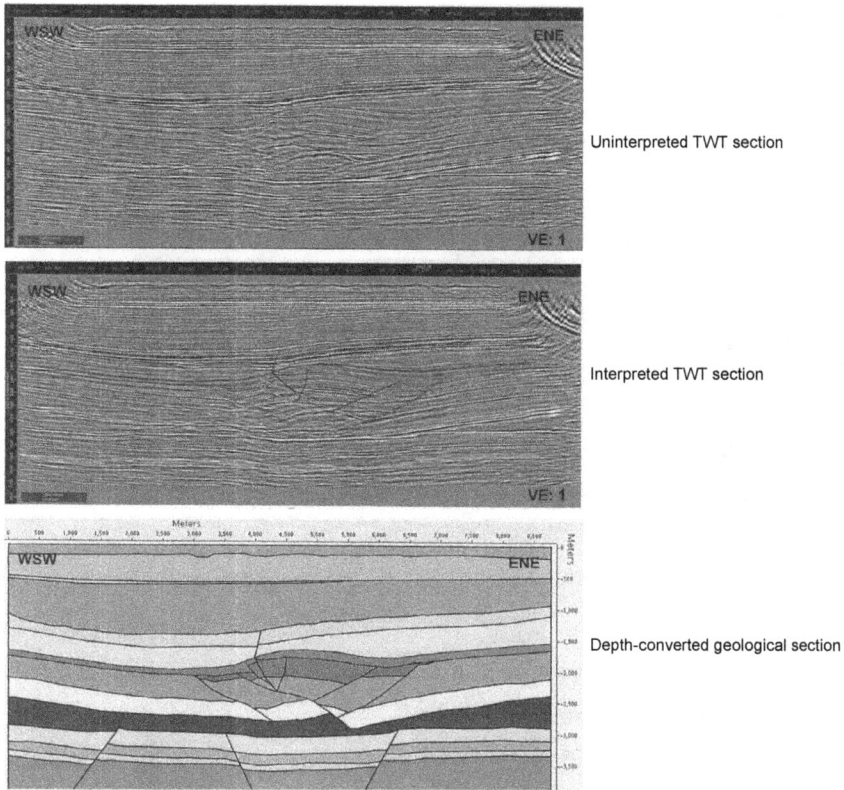

Figure 5.27. Example of the conversion of a seismic section into a geological cross-section, North Sea (after Robinson, 2014).

stratigraphy the strength, character and patterns of the main reflections are used to deduce the nature of the rocks themselves and their environment of deposition. An example of such a study is provided in Fig. 5.29, which shows a river channel, possibly leading to a delta, being mapped by 3D seismic interpretation.

Figure 5.30 shows an example from offshore Vietnam. It illustrates the development, from a seismic profile, of a depositional model, which in turn allows the reconstruction of the original environmental conditions — ranging from a coastal plain to shallow and deep marine — and finally the lithostratigraphic interpretation of the section.

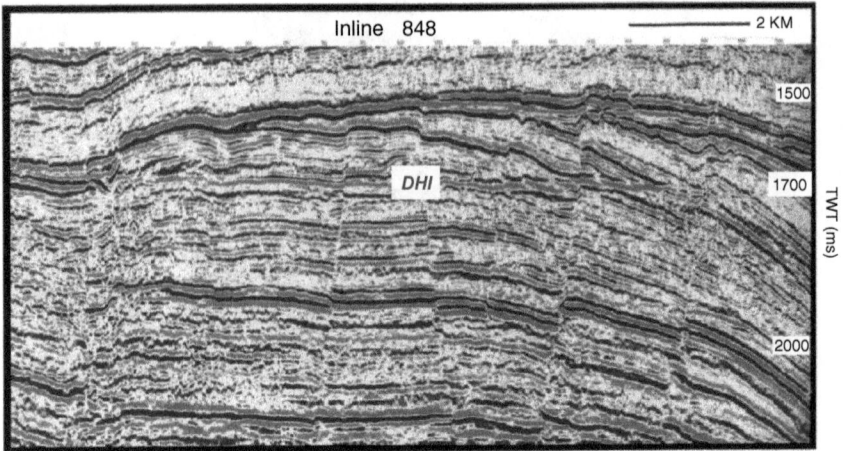

Figure 5.28. Seismic section across the Troll Gas Field, offshore Norway, showing extensive faulting and a 7-km-long flat spot at 1,700 ms, representing the GWC (Courtesy C. Jackson).

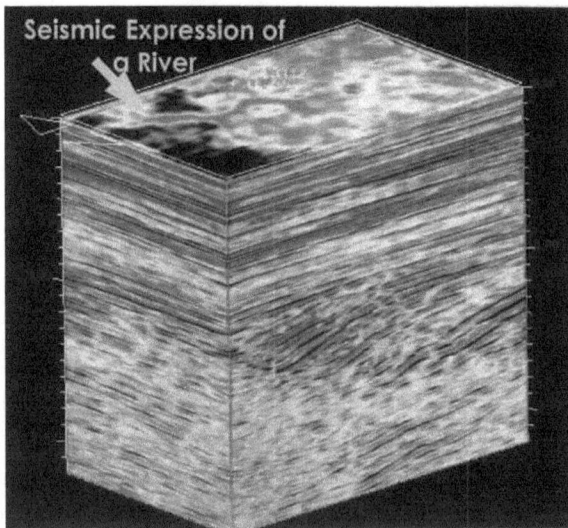

Figure 5.29. Mapping of a river channel by 3D seismic interpretation.

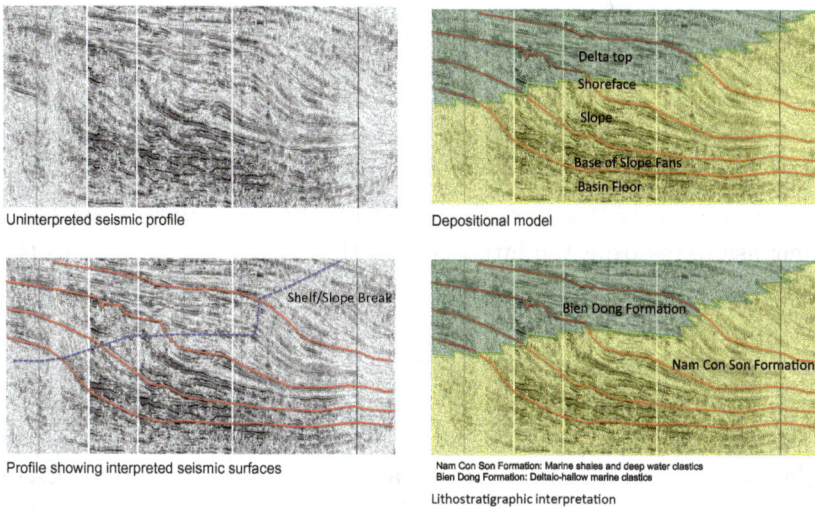

Uninterpreted seismic profile

Depositional model

Profile showing interpreted seismic surfaces

Lithostratigraphic interpretation

Figure 5.30. A seismic stratigraphic interpretation case study, offshore Vietnam (modified after Matthews *et al.*, 1997, courtesy A. Fraser).

Seismic stratigraphy has its own extensive terminology, the discussion of which is beyond the scope of this book. Its applications, however, may be summarised as follows:

- construction of depositional models,
- reconstruction of depositional environments,
- sedimentary facies analysis,
- geological history studies,
- prediction of rock properties such as lithology, bedding and porosity in undrilled areas,
- identification of sub-surface unconformities by the termination of reflections.

These are useful in guiding exploration programmes.

## 5.5.   Sub-surface Contour Maps

Contour maps are used extensively to show various characteristics of the rocks in the sub-surface and a *contour* is defined as a line that joins points of equal value.

### 5.5.1.  *Types of contour maps*

There is a great variety of contour maps but those most commonly used in petroleum geological studies are described briefly below.

#### 5.5.1.1.  *Structure contour maps*

*Structure contour maps* are the oldest types of sub-surface maps and their use dates from the latter half of the nineteenth century to the early years of the modern oil industry in the eastern USA. They can be constructed at field or basin scale and show the variation in the shape of a surface, often the contact between two different rock types, with respect to a datum. At field scale, such surfaces are normally the top and base of the reservoir; at basin scale they relate to the top of a laterally extensive rock unit, i.e. a formation or group. Sometimes the surface of interest is a regional unconformity. In most cases the datum is sea level.

Drilled depths to the surface of interest in oil and gas wells are the primary source of information. Structure contour maps are prepared by recording the depths with respect to sea level of the surface to be contoured on a base map at each control location — i.e. at each well — and drawing lines of equal depth through the data points. Steep dips are reflected in reduced spacing between adjacent contours, while widely spaced contours indicate gentle dips; closed contours indicate folds. Figures 5.31 and 5.32 show examples of structure contour maps.

Figure 5.31 shows two anticlines, with some faulting, separated by a syncline. The equal spacing between adjacent contours on the flanks means the folds are symmetrical. Figure 5.32 illustrates a number of folds with some faulting in the southeastern sector of the map. The reduction in spacing between adjacent contours reflects more steeply dipping northwestern anticlinal flanks than the southeastern ones, indicating that the folds are asymmetrical.

A structure contour map of the giant Abqaiq Oil Field, Saudi Arabia, is presented in Fig. 5.33. The contours are drawn on the top of the reservoir and show an elongated, north-northeast–south-southwest-trending, symmetrical anticline. The difference in

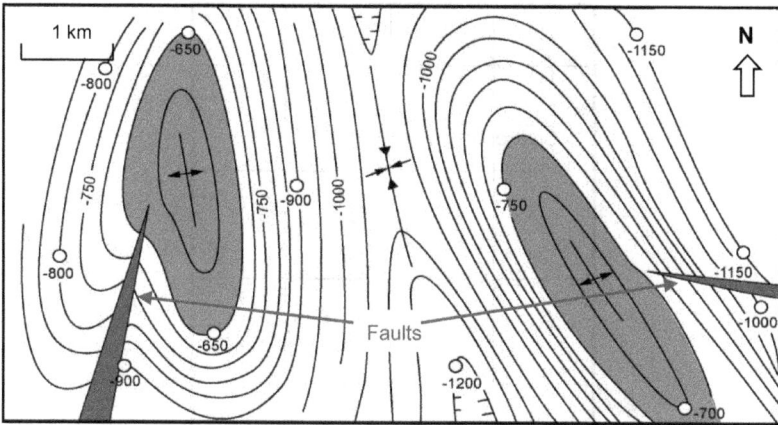

Figure 5.31. Two symmetrical anticlines separated by a syncline (contour intervals in metres).

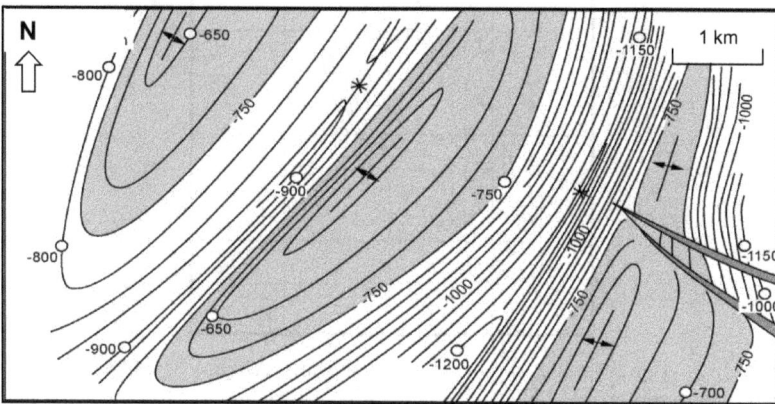

Figure 5.32. Asymmetrical anticlines and synclines (contour intervals in metres).

depth between the highest contour (5,800 ft) and the lowest closed contour (7,200 ft) represents the closure of the structure (the vertical distance between the crest of the fold and spill point; see Chapter 4, Fig. 5.37). In this case it amounts to 1,400 ft, which is substantial.

As a further example, a top reservoir structure contour map of the Karachaganak Oil and Gas-condensate Field, Kazakhstan, is shown

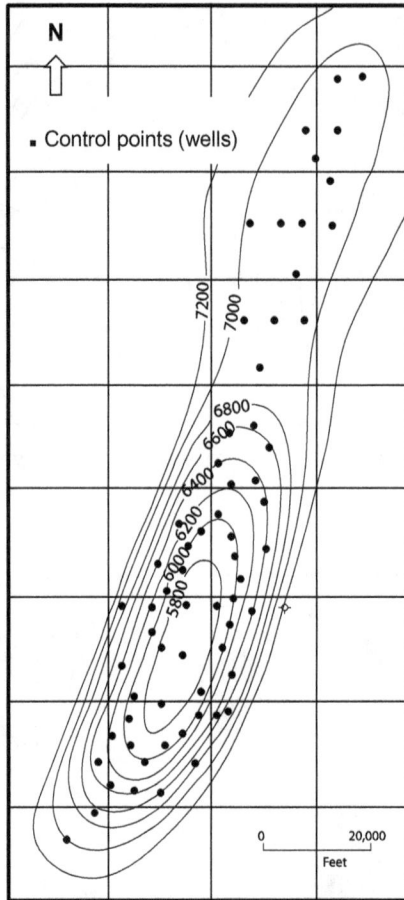

Figure 5.33. Top reservoir structure contour map, Abqaiq Field, Saudi Arabia, showing a symmetrical, NNE–SSW trending anticline with relatively gently dipping flanks. Closure: difference in depth between the highest contour and the lowest closed contour, 1,400 ft (modified and redrawn from Levorsen, 1967).

in Fig. 5.34. This is a dome-shaped structure and the OWC and GOC are plotted on the map, which is standard practice.

Field-scale structure contour maps play a vital part in the estimation of conventional reserves, since the areas enclosed by the GOC and OWC outline the oil- and gas-bearing parts of the trap.

Regional or basin-scale structure contour maps are prepared to show the variations in depth of a particular surface over large areas.

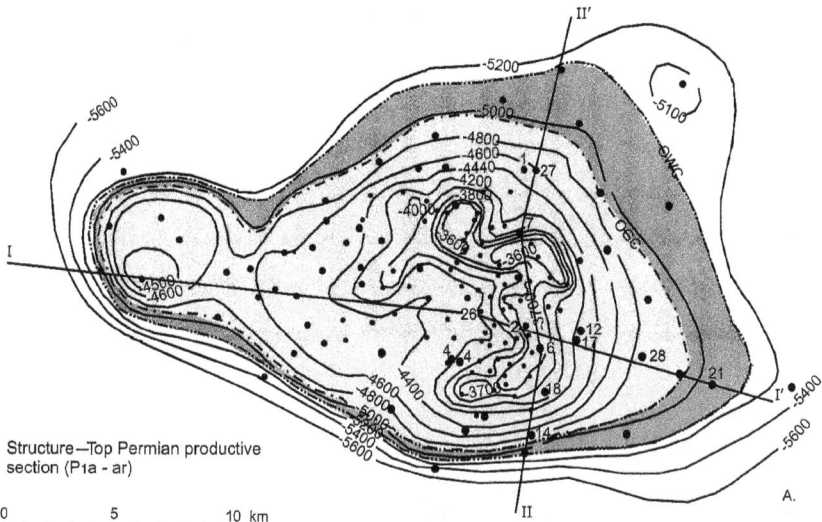

Figure 5.34. Top reservoir structure contour map, Karachaganak Oil and Gas-condensate Field, Kazakhstan, showing the fluid contacts (after Effimoff, 2001).

Often this is the top of a reservoir unit or a regional unconformity and such maps are useful guides to estimating depths to drilling targets. Figure 5.35 presents a structure contour map of the top of the Acacus Formation of Ordovician/Silurian age, a major play in the Ghadamis and Morzuk Basins in Libya.

### 5.5.1.2. *Isopach maps*

*Isopach maps* show the variations in the thickness of a specific interval and, like structure contour maps, can be constructed at both field and basin scales. They are constructed by plotting on a base map the thickness of the interval of interest at various control points, i.e. wells. Thickness values are obtained by subtracting the depth at which the top of the layer is encountered from the depth to its base in each well and then drawing lines of equal thickness through the data points.

At field scale, the interval of interest is the reservoir and an example of an isopach map of the producing zone at the Taipia Canyon Oil Field, California, is shown in Fig. 5.36. The reservoir reaches its maximum thickness of 120 ft in the northwestern sector

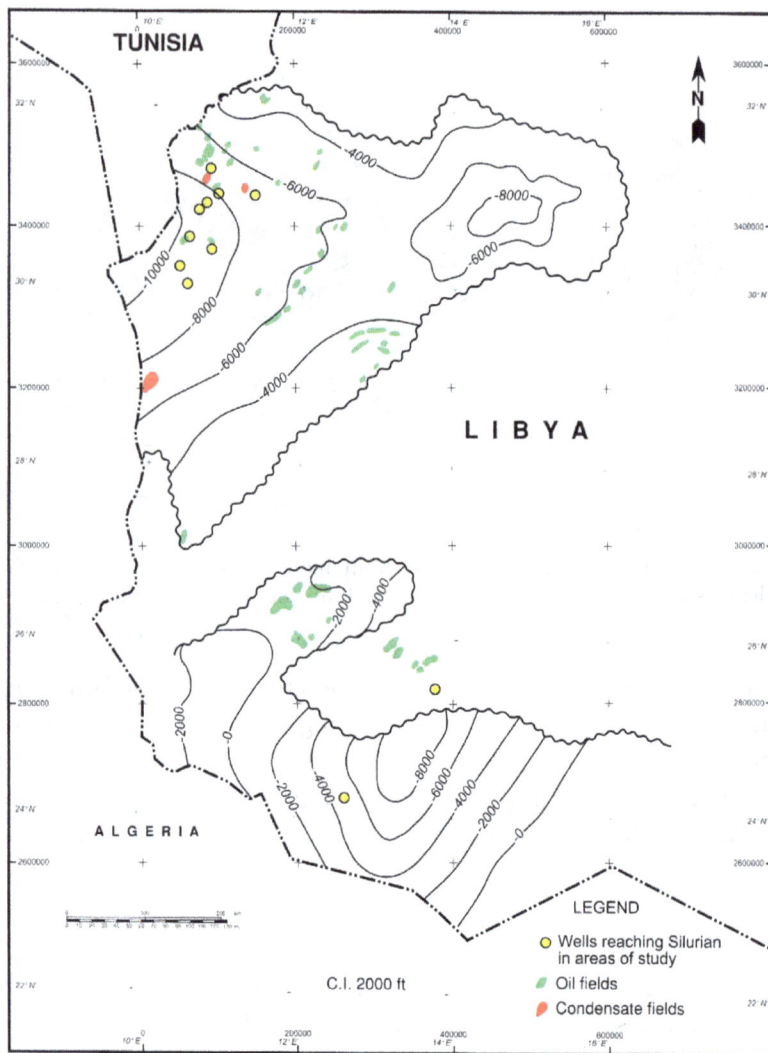

Figure 5.35. A regional top Acacus Formation structure contour map in the Ghadamis and Morzuk Basins, Libya (after Rusk, 2001).

of the field. Field-scale isopach maps can be used to determine the reservoir volume above hydrocarbon–water contacts and thus play an important part in the estimation of conventional reserves.

Regional or basin-scale isopach maps are prepared to show the thickness distribution of a particular layer or several layers over

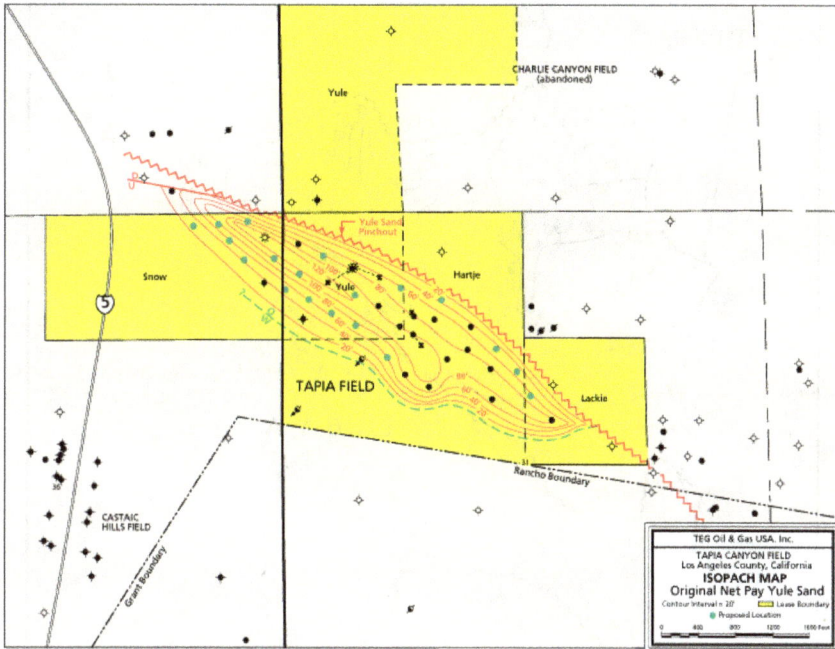

Figure 5.36. Isopach map of the reservoir at the Taipia Canyon Oil Field, northwest of Los Angeles, California (www.seftonresources.com).

large areas. This is commonly the reservoir, and such maps are useful in highlighting the areas of maximum reservoir development. Figure 5.37 presents an isopach map of the Acacus Formation in the Ghadamis and Morzuk Basins in Libya. The rock unit reaches its maximum thickness of 2,000 ft in the northwestern sector of the Ghadamis Basin.

### 5.5.1.3. *Facies maps*

Two- or several-component (multi-component) lithofacies maps are often prepared and are useful aids in assessing variations in reservoir quality and distribution as a function of lithofacies changes. A great variety of lithofacies maps exist but there is room for discussing only a few selected examples here. Figure 5.38 shows a three-component lithofacies map of the Middle Permian strata in the Permian Basin of Texas and New Mexico, where carbonates and

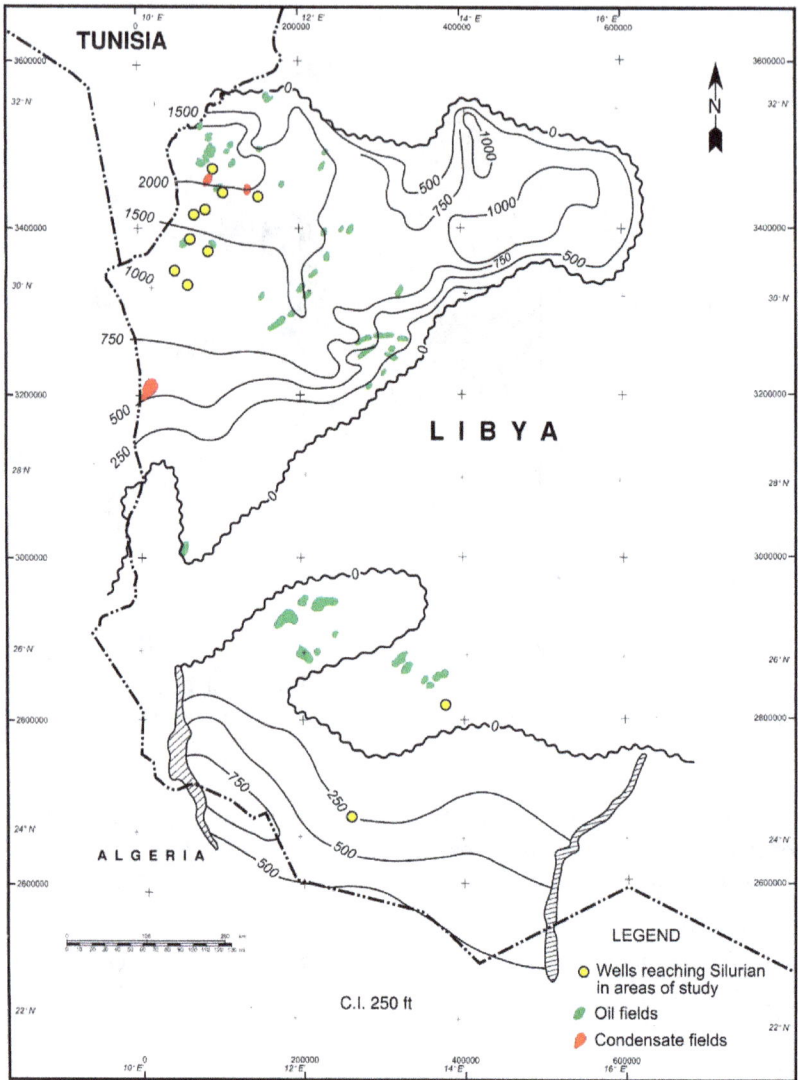

Figure 5.37. A regional isopach map of the Acacus Formation in the Ghadamis and Morzuk Basins, Libya (after Rusk, 2001).

sandstones form the reservoirs in the producing fields. A multi-component lithofacies map of the Lower Cretaceous strata in Kuwait and southern Iraq is presented in Fig. 5.39, where the sandstones host giant oil accumulations in several fields.

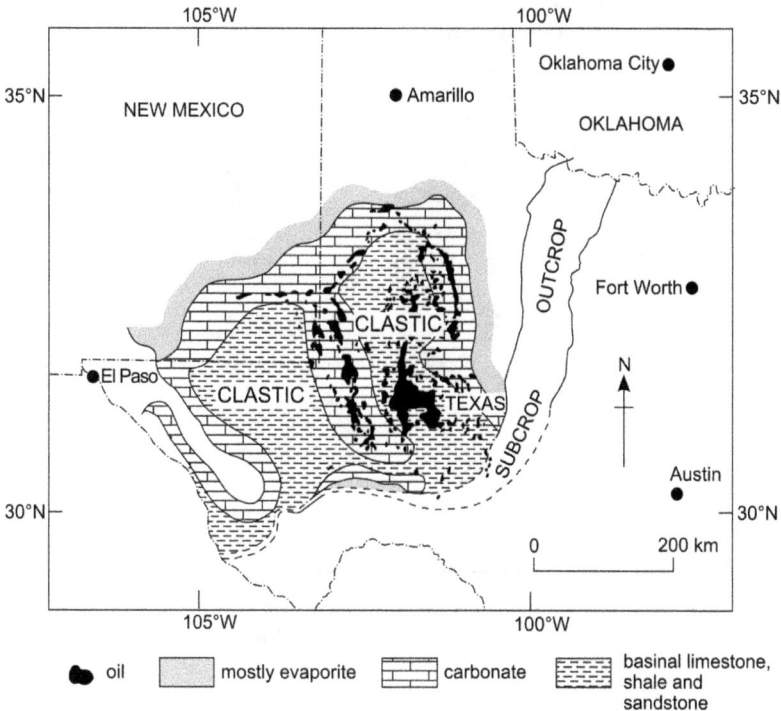

Figure 5.38. A three-component lithofacies map of the Middle Permian strata in the Permian Basin of western Texas and New Mexico (modified and redrawn from North, 1985).

The most commonly used types are the two-component sand–shale ratio and sandstone isolith maps. The former shows variations in the distribution of sand and shale over the area of interest. At each well, individual sand units are identified and their thicknesses added together. Similarly, the thicknesses of the shale layers are added together and the ratios of total sand to total shale thickness computed and recorded for each well on a base map. Lines of equal ratio are then drawn through the data points. Figure 5.40 shows a sand–shale ratio map of the Wilcox Group of Eocene age in the Gulf of Mexico coastal plain in Texas. The proportion of sand increases from the northwest to the southeast, indicating an improvement in the reservoir quality in the same direction.

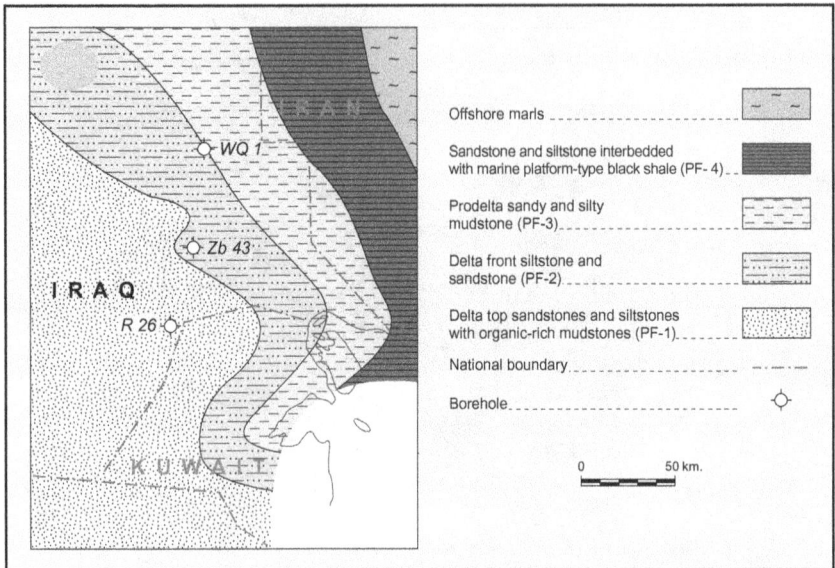

Figure 5.39. Multi-component lithofacies map of the Lower Cretaceous strata in Kuwait and southern Iraq (after Aqrawi *et al.*, 2010).

Isolith maps depict variations in the net thickness of a single rock type in the region of interest, the rock type invariably being sandstone. Hence they are called sand isolith maps. At each well, individual sand units are identified, their thicknesses totalled and these are plotted on a base map. Lines of equal thickness are then drawn through the data points.

A sand isolith map of the Wilcox Group is presented in Fig. 5.41. Areas of high isolith values coincide with those of high sand–shale ratios.

### 5.5.1.4. *Isoporosity and isopermeability maps*

*Isoporosity* and *isopermeability maps* respectively show lateral variations in mean porosity and mean permeability for a given interval. Examples of such maps for the Rough Gas Field, southern North Sea, are provided in Figs. 5.42 and 5.43. Both porosity and permeability are best developed in the central parts of the field and deteriorate progressively towards the southeast. Higher porosity indicates better

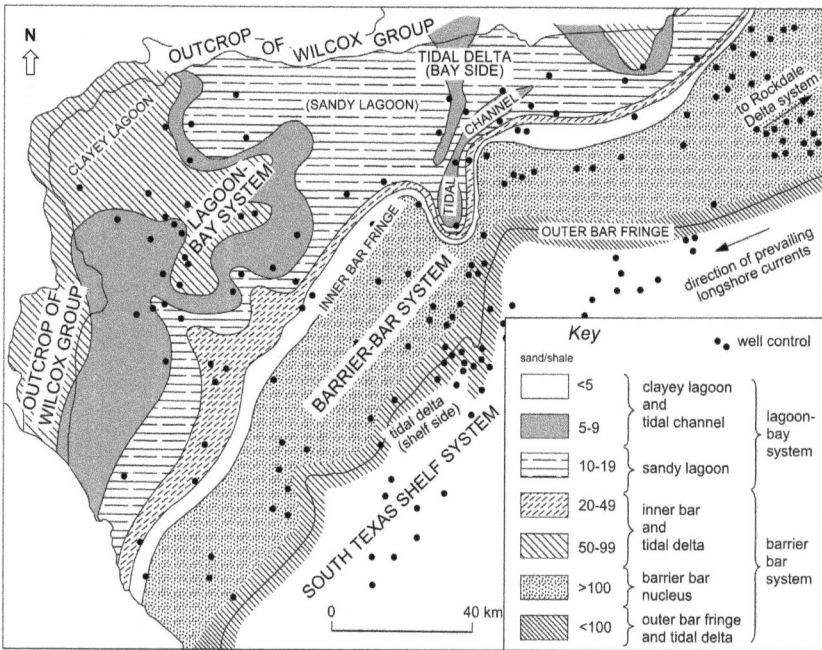

Figure 5.40. A sand–shale ratio map of the Eocene Wilcox Group in the Gulf of Mexico coastal plain in Texas, showing an increase in sand content from the northwest to the southeast (modified and redrawn from North, 1985).

reservoir quality and high flow rates are associated with high permeability values. Consequently, the production capacity of the reservoir is best in the central parts of the field.

## 5.6.   Leads, Prospects and Risk Analysis

Exploration is a risky venture and only a fraction of the wells drilled are successful, even in established petroliferous provinces. Following the preliminary stages of their investigations, explorationists identify *leads* and *prospects* and rank them in terms of their hydrocarbon potential. A lead is defined as a possible target for exploration drilling, identified on the basis of regional geological study and/or 2D seismic data interpretation. It requires further evaluation before drilling. A prospect is defined as a firm drilling target, identified through detailed evaluation and, normally, 3D seismic data interpretation.

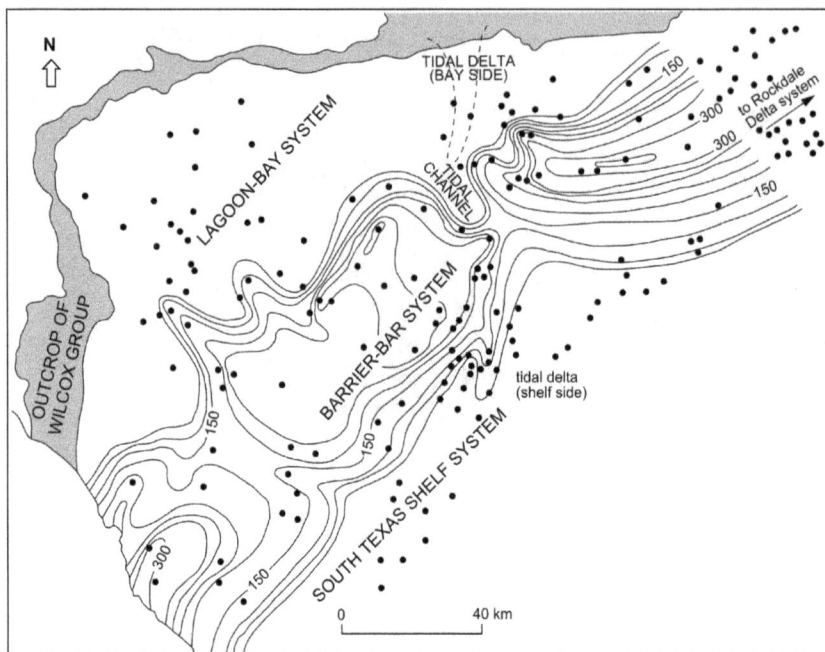

Figure 5.41. A sand isolith map of the Eocene Wilcox Group in the Gulf of Mexico coastal plain in Texas, showing coincidence between areas of high isolith values and high sand–shale ratios (modified and redrawn from North, 1985).

The next step is the construction of a *prospect map*, showing the locations of the areas where hydrocarbon generation is expected to have taken place and where potential traps are present. The former are referred to as *kitchens* and generally coincide with the central parts of basins where the sediments have been buried to great depths and subjected to high temperatures, resulting in source-rock maturation and hydrocarbon generation. Once generated, the hydrocarbons will move up the basin flanks and fill any traps that may be present along their migration paths. Such traps would represent prime drilling targets. Figure 5.44 presents a prospect map of an offshore area southwest of Greenland. It outlines two potential kitchens, indicates the possible hydrocarbon migration routes and identifies several traps with four-way dip closure that lie in the path of the migrating hydrocarbons.

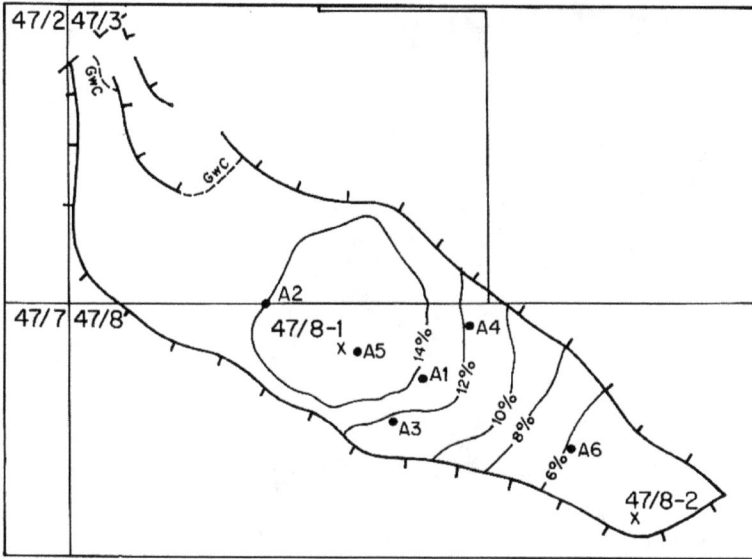

Figure 5.42. Isoporosity map, Rough Gas Field, southern North Sea (from Archer and Wall, 1986).

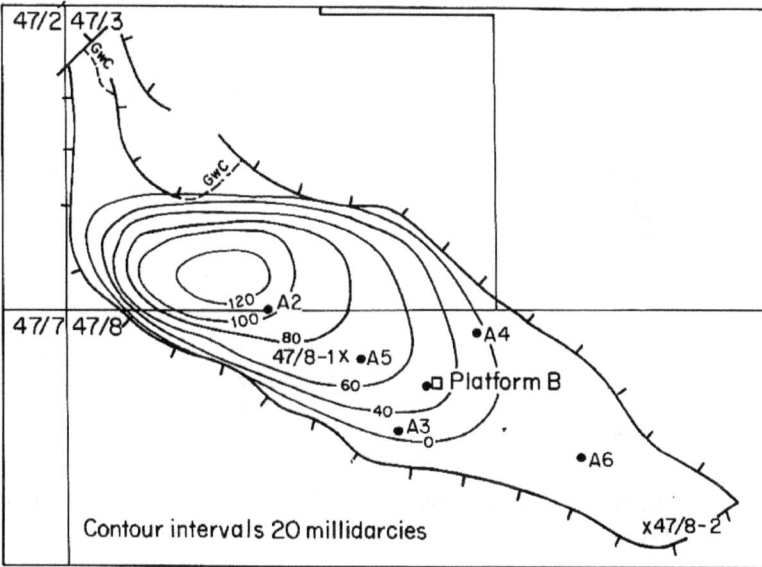

Figure 5.43. Isopermeability map, Rough Gas Field, southern North Sea (from Archer and Wall, 1986).

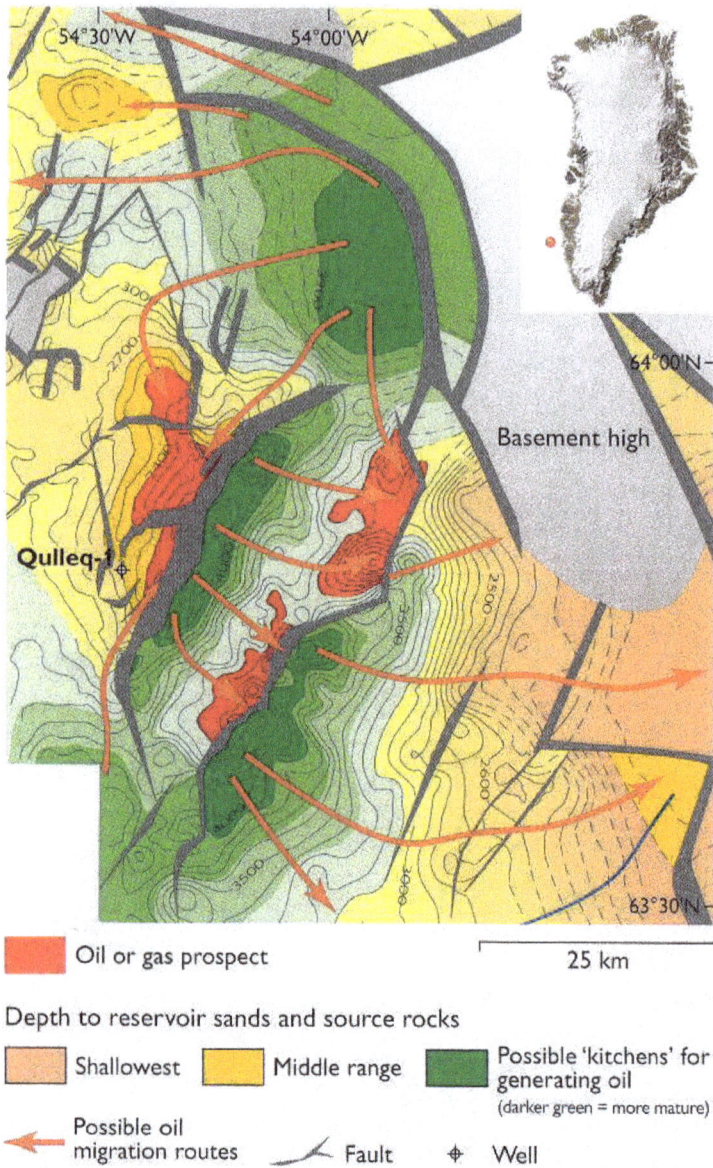

Figure 5.44. An example of a prospect map, offshore southwest Greenland (after Henriksen, 2008).

## 5.6.1.  *Risk analysis*

As discussed in some detail in Chapter 4, the geological conditions controlling petroleum occurrence are:

- mature source rocks,
- presence of reservoir rocks,
- presence of traps,
- presence of cap rocks or seals,
- favourable timing of trap formation in relation to petroleum migration.

In assessing the geological chance of success or risk, a probability, $p$, is assigned to each parameter being fulfilled:

$p_1$ = probability of presence of mature source rocks,
$p_2$ = probability of presence of reservoir rocks,
$p_3$ = probability of presence of traps,
$p_4$ = probability of presence of seals,
$p_5$ = probability of favourable timing.

A value between 0 and 1 is assigned to $p_1, p_2, p_3, p_4$ *and* $p_5$ and the overall probability or level of risk is determined from their product:

$$\text{probability of finding oil or gas} = p_1 \times p_2 \times p_3 \times p_4 \times p_5. \quad (5.2)$$

If several areas or individual prospects are being considered, they may be analysed one by one, placed in a matrix, their probability of success computed and an order of merit established, as illustrated in Fig. 5.45.

The risk can then be categorised as shown in Fig. 5.46.

With a geological chance of success rating of 0.504, prospect A falls in the low-risk category and would be the most attractive drilling target. It should be noted, however, that the above is an analysis of technical risk. In addition to these purely geological parameters, other considerations such as economic and political

**PROSPECTS**

|  | A | B | C | D |
|---|---|---|---|---|
| Source probability | 0.8 | 0.6 | 0.7 | 0.4 |
| Reservoir probability | 0.9 | 0.8 | 0.6 | 0.3 |
| Trap probability | 1.0 | 0.7 | 0.5 | 0.5 |
| Seal probability | 0.7 | 0.8 | 0.7 | 0.2 |
| Favourable timing | 1.0 | 1.0 | 1.0 | 1.0 |
| *Probability of success* | *0.504* | *0.269* | *0.147* | *0.012* |

Figure 5.45. Prospect risk analysis.

| RISK LEVEL | AVERAGE PROBABILITY OF GEOLOGICAL SUCCESS $(p_1 \times p_2 \times p_3 \times p_4 \times p_5)$ |
|---|---|
| Very low | 0.750 |
| Low | 0.375 |
| Moderate | 0.183 |
| High | 0.092 |
| Very high | 0.050 |

Figure 5.46. Categorisation of risk.

factors, as well as proximity to infrastructure such as pipelines and loading terminals, must also be taken into account.

## 5.7.  Play Fairway Analysis (PFA)

The above discussion relates to the assessment of exploration risk at a prospect level. More recently, the concept of risk analysis has been

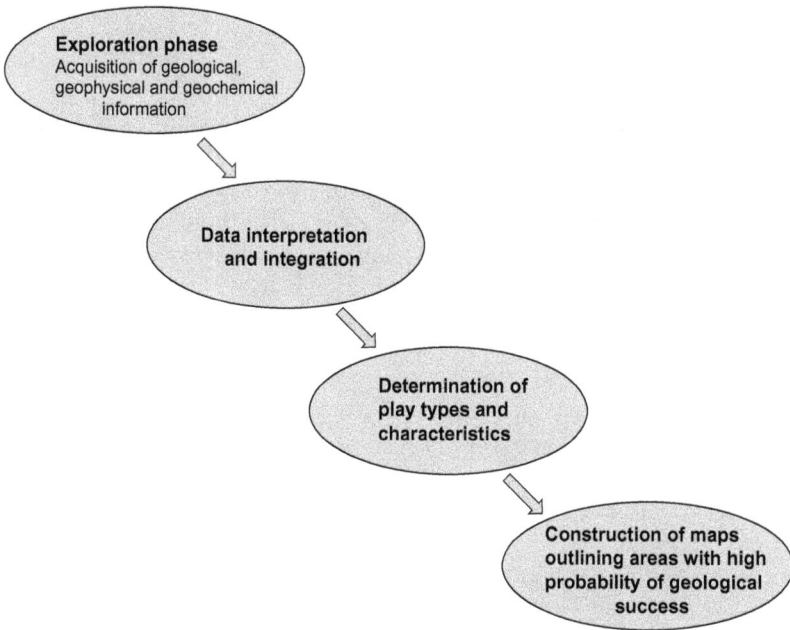

Figure 5.47. Steps involved in play fairway analysis.

extended to identifying the regional distribution of the key elements of hydrocarbon plays. PFA is thus essentially an assessment of exploration risk at a basin or regional scale. It involves integrating the geological and geophysical information collected during the preceding exploration phase as indicated in the simplified flow chart presented in Fig. 5.47.

This is particularly useful in frontier basins (underexplored or unexplored areas), since it allows exploration effort to be concentrated in the most prospective parts of such basins. Furthermore, where several basins are under consideration, PFA provides a means of ranking them, allowing exploration to be focused on particular basins. In more mature areas, the technique can be applied to identifying new plays in underexplored parts of the basin or provide an indication that the basin has little remaining potential and that it may be appropriate to contemplate relinquishing the area.

## References

Aqrawi, A.M.A., Goff, J.C., Horbury, A.D. *et al.* (2010). *The Petroleum Geology of Iraq*, Scientific Press, UK.

Archer, J.S. and Wall, C.G. (1986). *Petroleum Engineering: Principles and Practice*, Graham and Trotman, UK.

Bjorlykke, K. (2010). *Petroleum Geoscience: From Sedimentary Environments to Rock Physics*, Springer, London, UK.

Bureau of Minerals and Petroleum, Greenland (2009). *Annual Report, 2008*, BMP Greenland, Nuuk, Greenland.

Effimoff, I. (2001). "Future hydrocarbon potential of Kazakhstan", in Downey, M.W., Threet, J.C. and Morgan, W.A. (eds.), *Petroleum Provinces of the Twenty-First Century, AAPG Memoir 74*, AAPG, Tulsa, OK.

Fox, A.F. (1970). "General geological introduction", in *Our Industry Petroleum*, British Petroleum Company Ltd., London, UK.

Global Geo Services AS (2000). The Persian Carpet Seismic Survey, www.GGS no.

Henriksen, N. (2008). *Geological History of Greenland*, Geological Survey of Denmark and Greenland (GEUS), Copenhagen, Denmark.

Levorsen, A.I. (1967). *Geology of Petroleum, 2nd Edition*, W.H. Freeman, San Francisco, CA.

Lowell, J.D. (1985). *Structural Styles in Petroleum Exploration*, Pennwell Corp., Tulsa, OK.

Marshak, S. (2005). *Earth: Portrait of a Planet, 2nd Edition*, W.W. Norton, New York, NY.

Matthews, S.J., Fraser, A.J., Lowe, S.P. *et al.* (1997). "Structure, stratigraphy and petroleum geology of the SE Nam Con Son Basin, offshore Vietnam", in Fraser, A.J., Matthews, S.J and Murphy, R.W. (eds.), *Petroleum Geology of Southeast Asia, Geological Society Special Publication 126*, The Geological Society Publishing House, Bath, UK.

Morzuk Basin, Libya, in 'Sand Sea in Southwestern Libya (2011)'. Available online at: https://earthobservatory.nasa.gov/IOTD/view.php?id=76652.

North, F.K. (1985). *Petroleum Geology*, Allen and Unwin, London, UK.

Pacheco, M., Martin-Barajas, A., Elders, W. *et al.* (2006). Stratigraphy and structure of the Altar Basin of NW Sonora: Implications for the history of the Colorado River Delta and the Salton Trough, *Revista Mexicana de Ciencias Geológicas*, **23**, 1–22.

Piggott, H.D.G. (1977). "Geophysical prospecting", in *Our Industry Petroleum*, British Petroleum Company Ltd., London, UK.

Robinson, S. (2014). 3D structure of the Kelvin Collapse Graben, North Sea, unpublished MSc dissertation, Imperial College London, UK.

Rusk, D.C. (2001). "Libya: Petroleum potential of the underexplored Basin centres — a Twenty-First-Century Challenge", in Downey, M.W., Threet, J.C. and Morgan, W.A. (eds.), *Petroleum Provinces of the Twenty-First Century, AAPG Memoir 74*, AAPG, Tulsa, OK.

# Chapter 6

# Resources and Reserves: Definition, Classification and Quantification

## 6.1. Introduction

The term *resource* has a broad meaning and encompasses accumulations of all industrially useful materials, which includes mineral deposits as well as petroleum. *Reserves* refer to the portion of an identified resource that can be produced economically through existing technology. As far as petroleum is concerned, a distinction is normally made between *conventional* and *unconventional* oil and gas. Conventional hydrocarbons occur in pores in the reservoir rock from which they are extracted by drilling production wells. Unconventional hydrocarbons, by contrast, are associated with tight shale rocks, which must be artificially fractured in order to enable the oil and gas to flow at economically viable rates. Until relatively recently, commercial production of hydrocarbons from shale was not possible but a technical breakthrough in the US in the mid-1990s dramatically changed this situation. The breakthrough was the introduction of new technology enabling wells to be drilled horizontally over distances of up to 10−12 km through shale layers and combining this with hydraulic fracturing, abbreviated nowadays to *fracking*. Oil and gas supplies from shale have risen rapidly in the US since the turn of the present century and have transformed the American energy landscape; the US is now self-sufficient in gas and its imports of crude oil have diminished significantly. It should be noted, however, that although hydrocarbon-bearing shale rocks are present in many parts of the world, large-scale oil and gas production from shale is a North American phenomenon and is yet to be replicated elsewhere.

For the sake of completeness, mention should also be made of heavy, viscous oil deposits that are present at or close to the surface in several parts of the world. The largest of these are in Canada and Venezuela and consist of sands impregnated with 5–15° API gravity oil. Known as tar or oil sands, the deposits contain vast amounts of oil and were traditionally placed in the unconventional category, together with shale gas and shale oil, and regarded as resources. However, advances in extraction technology have made it possible for such hydrocarbons to be produced economically, allowing shale gas, shale oil and oil sand resources to be classified as reserves.

## 6.2. Classification and Quantification of Reserves

In 2011 a number of organisations[a] jointly produced a comprehensive document entitled *Guidelines for Application of the Petroleum Resources Management System,* the aim of which was to establish guidelines and standardise procedures for the assessment and classification of oil and gas resources. The guidelines are followed by the international petroleum industry since they provide a common reference and improve clarity in global communications relating to petroleum resources.

The definitions of resources, reserves and the associated terminology are explained Fig. 6.1. Reserves are subdivided into several categories, which depend on the degree of confidence in the computed value of the estimate. These estimates are discussed briefly below.

- *Hydrocarbons in place* (also referred to as *in place reserves*): This represents the total volume of oil or gas that is present in a field

---

[a]Society of Petroleum Engineers (SPE), American Association of Petroleum Geologists (AAPG), World Petroleum Council (WPC) and Society of Petroleum Evaluation Engineers (SPEE). The group is referred to by the abbreviated term SPE-PRMS.

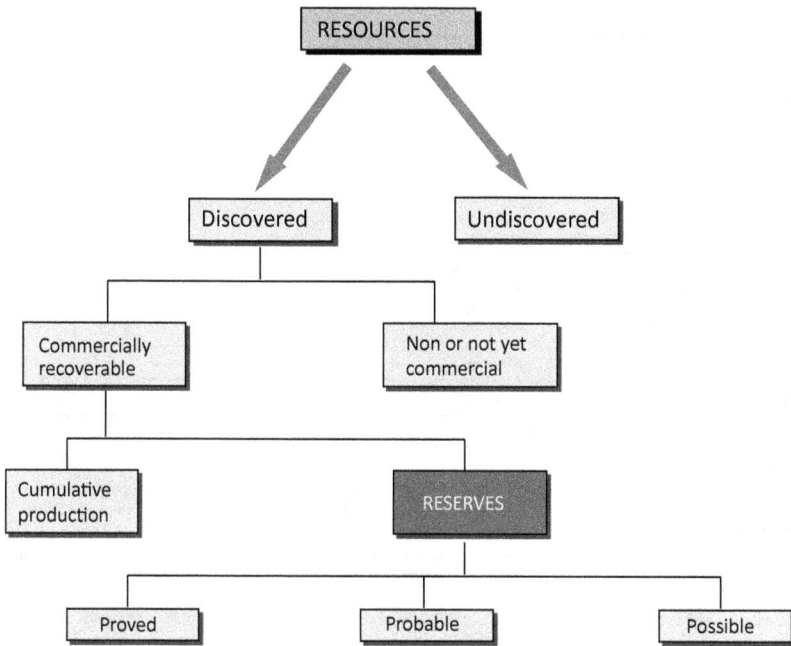

Figure 6.1. Definition of resources and reserves.

before any production and is referred to as *stock tank oil initially in place*, abbreviated to *STOIIP*. This quantity is not all recoverable.

- *Proved reserves*: Economically, this is the most important category and refers to the oil or gas in existing fields that can be produced from existing wells.
- *Probable reserves*: This category includes oil or gas in extensions of existing fields beyond or below the currently known limits.
- *Possible reserves*: This is oil or gas expected from future discoveries in areas or formations known to be productive.
- *Undiscovered reserves*: This category is subject to the greatest degree of uncertainty and refers to the quantity of oil and gas that could be found in an area.

Figure 6.2 illustrates the resource classification framework that is currently in use in the petroleum industry.

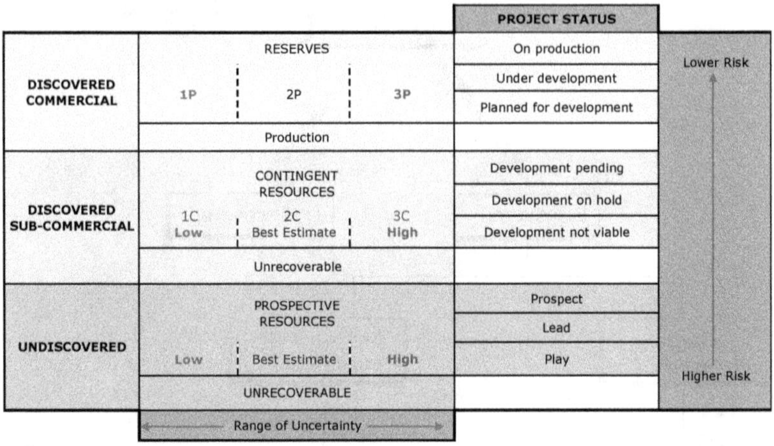

| | | | | PROJECT STATUS | |
|---|---|---|---|---|---|
| | | RESERVES | | On production | Lower Risk |
| DISCOVERED COMMERCIAL | 1P | 2P | 3P | Under development | |
| | | | | Planned for development | |
| | | Production | | | |
| DISCOVERED SUB-COMMERCIAL | 1C Low | CONTINGENT RESOURCES 2C Best Estimate | 3C High | Development pending | |
| | | | | Development on hold | |
| | | | | Development not viable | |
| | | Unrecoverable | | | |
| UNDISCOVERED | Low | PROSPECTIVE RESOURCES Best Estimate | High | Prospect | |
| | | | | Lead | |
| | | | | Play | Higher Risk |
| | | UNRECOVERABLE | | | |
| | ←————— Range of Uncertainty —————→ | | | | |

**Definitions:**

*Contingent resources*: those quantities of petroleum estimated, as of a given date, to be potentially recoverable from known accumulations but which are not currently commercially recoverable due to lack of a viable market, or where economic recovery is dependent on the development of new technology or where evaluation is at an early stage.

*1P*: equivalent to proved reserves; denotes low estimate scenario of reserves; *2P*: equivalent to the sum of proved plus probable reserves; denotes best estimate scenario of reserves; *3P*: equivalent to the sum of proved plus probable plus possible reserves; denotes high estimate scenario of reserves.

*1C*: denotes low estimate scenario of contingent resources; *2C*: denotes best estimate scenario of contingent resources; *3C*: denotes high estimate scenario of contingent resources.

Figure 6.2. Resource classification framework (modified from SPE-PRMS, 2011).

## 6.3.  Reserve Estimation

Broadly, reserve estimation methods fall into two groups: (a) volumetric and (b) performance analysis. The latter method is complex, requires relatively long-term production data, which are not available at the time of, or shortly after, a discovery, and is beyond the scope of this discussion. This section will therefore concentrate on the volumetric method, which is based on two different techniques, namely the *deterministic* and *probabilistic* or *stochastic* approaches.

### 6.3.1.  *The deterministic approach*

The *deterministic approach* leads to a single value for the reserves or STOIIP. The computation can be performed on a pocket calculator and involves selecting the best available average values of the various parameters that are used in the STOIIP equation. Outlined below is the procedure that is normally followed.

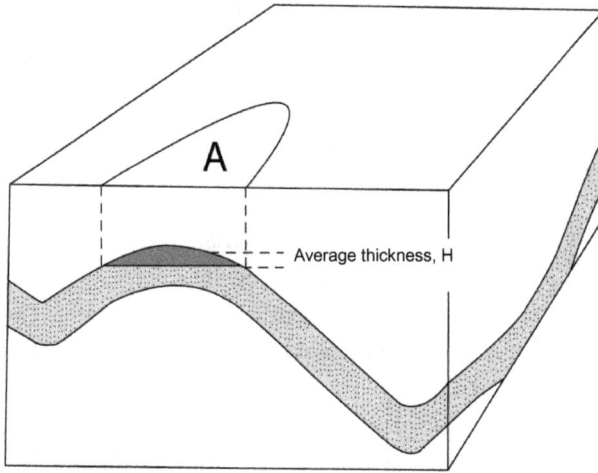

GRV = A x average thickness
A = hydrocarbon-bearing area above the hydrocarbon/water contact (ft or m)
H = gross reservoir mean thickness above the hydrocarbon/water contact (ft or m)

Figure 6.3. Determination of gross rock volume (modified and redrawn from White and Gehman, 1979).

The first step is to ascertain the volume of the potentially hydrocarbon-bearing rock. This quantity is the product of the area of the accumulation (determined from structure contour maps) and the mean gross reservoir thickness above the hydrocarbon/water contact. It represents the gross rock volume (GRV) of the potentially productive reservoir, as illustrated in Fig. 6.3:

$$\text{GRV} = A \times H, \tag{6.1}$$

where, $A$ is the productive area (acres or hectares) and $H$ is the gross reservoir mean thickness above the hydrocarbon/water contact (ft or m).

Oil volumes are reported either in cubic metres or barrels. Gas reserves are reported in cubic metres or cubic feet and are sometimes expressed in terms of their barrels of oil equivalent (BoE). Table 6.1 lists the conversion factors used in the petroleum industry.

Reservoirs invariably contain some non-productive intervals (layers with little or no porosity), which do not contribute to the storage

Table 6.1. Conversion factors.

| | |
|---|---|
| 1 barrel = 42 US gallons | 1 tonne = 7 barrels (approximately) |
| = 35 imperial gallons | 1 m$^3$ = 6.3 barrels |
| = 159 litres | = 37 ft$^3$ |
| 1 barrel is equivalent to 6,000 ft$^3$ of gas for conversion purposes (BoE) | |

capacity of the rock. They must be discounted, which means that distinction should be made between the gross reservoir interval and the *net pay* section. The term *pay* means that the interval is hydrocarbon-bearing and net pay represents the total thickness of the permeable intervals above the hydrocarbon–water contact. Impermeable zones and hydrocarbon-bearing intervals can be identified from the measurements of their radioactive and electrical properties by running logs in wells (see Chapter 8). The gamma ray log is normally, but not always, an indicator of the impermeable zones, and electrical resistivity logs are useful in the identification of the pore fluids; being non-conductive, hydrocarbon-bearing intervals are associated with high-resistivity log responses. To obtain the net pay volume of the reservoir, GRV is multiplied by a factor known as the *net-to-gross ratio*, symbolised by $N/G$. Figure 6.4 illustrates how $N/G$ can be computed and explains the terminology associated with the determination.

Selecting the best available average values of porosity and hydrocarbon saturation of the net pay section, the magnitude of the hydrocarbons in place may be estimated. For liquid hydrocarbons, this quantity is referred to as the *oil initially in place*, abbreviated to OIIP:

$$\text{OIIP} = \text{GRV} \times N/G \times \Phi(1 - S_w), \tag{6.2}$$

where $\Phi$ is the best average porosity value (fraction), $S_w$ is the best average water saturation value (fraction), $1 - S_w$ represents the hydrocarbon saturation.

OIIP can be converted into reserves in terms of stock tank oil initially in place, STOIIP, by applying an additional factor:

$$\text{STOIIP} = [\text{GRV} \times N/G \times \Phi(1 - S_w)]/B_o. \tag{6.3}$$

Gamma Ray          Resistivity

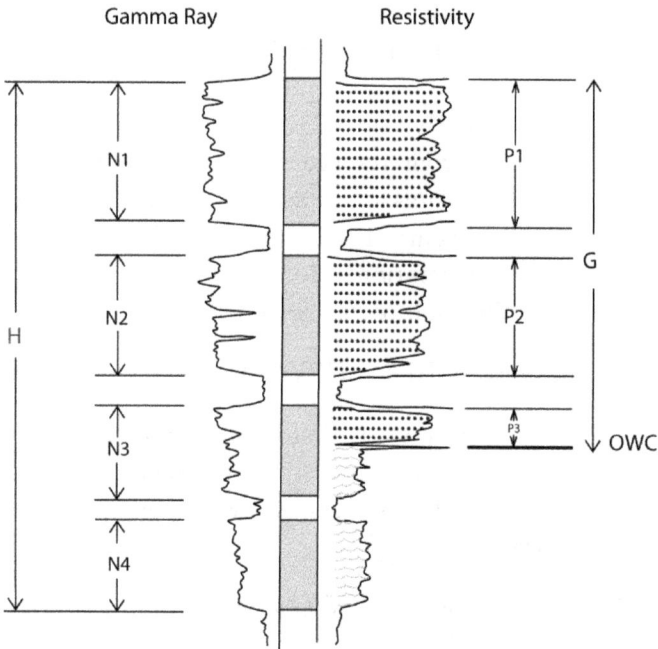

**Definitions:**

*H* = gross reservoir thickness

*G* = gross pay thickness

N1 + N2 + N3 + N4 = net reservoir thickness

P1 + P2 + P3 = net pay thickness (reservoir rock above the OWC. Lower part of
N3 is excluded)

Net-to-gross ratio, *N/G* = (P1 + P2 + P3)/G

Figure 6.4. Determination of $N/G$ and illustration of the associated terminology (modified and redrawn from Dikkers, 1985).

$B_o$ is called the *oil formation volume factor*. This parameter is essential for converting oil volumes from reservoir to stock tank conditions. At reservoir pressure and temperature, the oil contains gas in solution. This gas comes out of solution as the oil reaches the storage tanks, where pressure and temperature drop to surface conditions, causing a shrinkage in the volume of the oil. A reservoir barrel, therefore, occupies more space than its surface counterpart. The magnitude of the shrinkage depends on a parameter known the *gas–oil ratio* of the crude. It is abbreviated to GOR and defined as the volume of gas, measured in cu ft, released by a barrel of the oil as it reaches the surface. $B_o$ is obtained from laboratory tests, referred to as pressure–volume–temperature (PVT) analysis and its most common range is $1.1 - 1.6$. Its value rises with increasing GOR values.

Gas responds differently than oil to pressure reduction; it expands as the pressure drops from reservoir to surface conditions, which means that a cubic foot of gas at surface pressure occupies more space than its reservoir counterpart. The increase is in fact several-fold and, consequently, the gas formation volume factor, $B_g$, is applied quite differently in the estimation of gas reserves:

$$\text{Gas initially in place (GIIP)} = \text{GRV} \times N/G \times \Phi(1 - S_w) \times B_g. \quad (6.4)$$

As mentioned earlier, only a fraction of the hydrocarbon quantities originally present in a reservoir is recoverable. Application of a *recovery factor*, $R$, is therefore necessary to convert STOIIP into recoverable reserves:

$$\text{Oil reserves } = [\text{GRV} \times N/G \times \Phi(1 - S_w) \times R]/B_o, \quad (6.5)$$

$$\text{Gas reserves } = \text{GRV} \times N/G \times \Phi(1 - S_w) \times R \times B_g. \quad (6.6)$$

$R$ depends on a number of parameters, including the production mechanism, permeability, rock/fluid interfacial tension and hydrocarbon viscosity and density. The production mechanism, the natural process driving the hydrocarbons through and out of the reservoir into the production wells, has a major impact on recovery efficiency. Production or drive mechanisms fall into three principal types: (a) solution gas drive, (b) gas cap expansion drive, and (c) water drive.

In a reservoir under solution gas drive, as the pressure declines, gas bubbles out of the oil, expands and forces the oil out of the pores, towards and up the wellbores to the surface. At the same time, the expansion of the gas helps to maintain reservoir pressure.

Figure 6.5 illustrates the mechanism of gas cap expansion drive, which operates in accumulations containing both oil- and gas-bearing zones. Also, the gas coming out of solution in the oil leg moves up and joins the existing gas cap, which expands as the reservoir pressure drops due to production. This causes the GOC to move downwards,

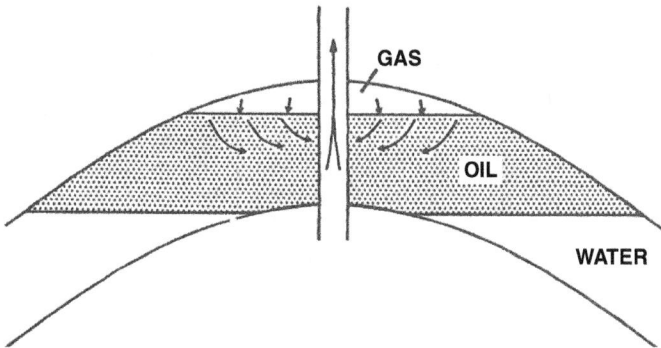

Figure 6.5. Illustration of the gas cap expansion drive mechanism. GOC moves down and gas fills the pore spaces vacated by oil due to production (after Stoneley, 1995).

pushing the oil through the reservoir into and up the production wells to the surface.

In a water drive, the water from below the OWC moves up and occupies the pores vacated by oil as the result of production. The OWC is pushed upwards, forcing the oil through the reservoir into the production wells to the surface. It should be noted, however, that a water drive becomes effective only if the reservoir is regionally extensive and connected to its outcrops, which enables surface waters to percolate through the rock to ensure a continuous supply of water to replace the oil that is being produced, as illustrated in Fig. 6.6. The mechanism of water drive is shown in Fig. 6.7.

Water drive is the most efficient production mechanism; solution gas drive is the least efficient and gas cap expansion is somewhere in between. The recovery factor is thus a highly variable parameter; it can be as high as 50% or 60% in clean (low silt and clay content), oil-bearing sandstone reservoirs with good inter-granular porosity and under an active water drive, and as low as 15% or 20% in complex carbonate reservoirs with fracture or vuggy porosity and under solution gas or gas cap expansion drive. Recovery from gas-bearing reservoirs can be as high as 90% due to the low viscosity and density of gas.

It is interesting to note that the hydrocarbons initially in place constitute only a tiny fraction of the gross reservoir volume. As

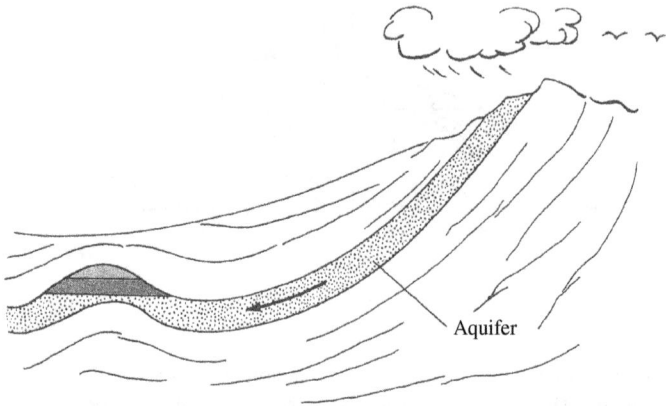

Figure 6.6. Geological setting for an active water drive mechanism. Water from the aquifer replaces the oil that is being produced (after Selley, 1997).

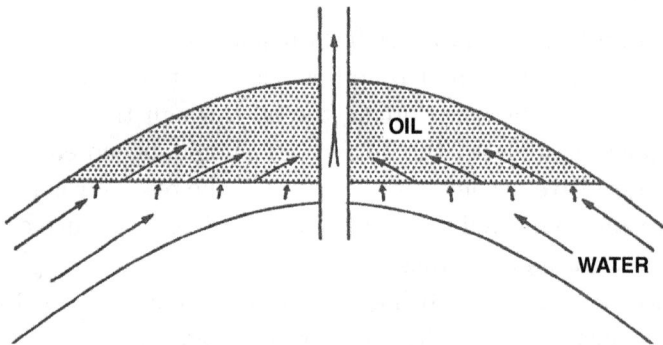

Figure 6.7. Illustration of the water drive mechanism. OWC moves up and gas fills the pore spaces vacated by oil due to production (after Stoneley, 1995).

illustrated in Fig. 6.8, starting with a GRV of 100 volume units and assuming $N/G = 0.8$, $\Phi = 0.25$, $1 - S_W$ (hydrocarbon saturation) $= 0.8$ and $B_o = 1.3$, STOIIP will amount only to 13 volume units. Application of a recovery factor will reduce this even further.

### 6.3.2.  *The probabilistic approach*

A fairly high degree of uncertainty is associated with all the parameters in the STOIIP equation used in deterministic reserves estimation. These uncertainties obviously affect the calculated values

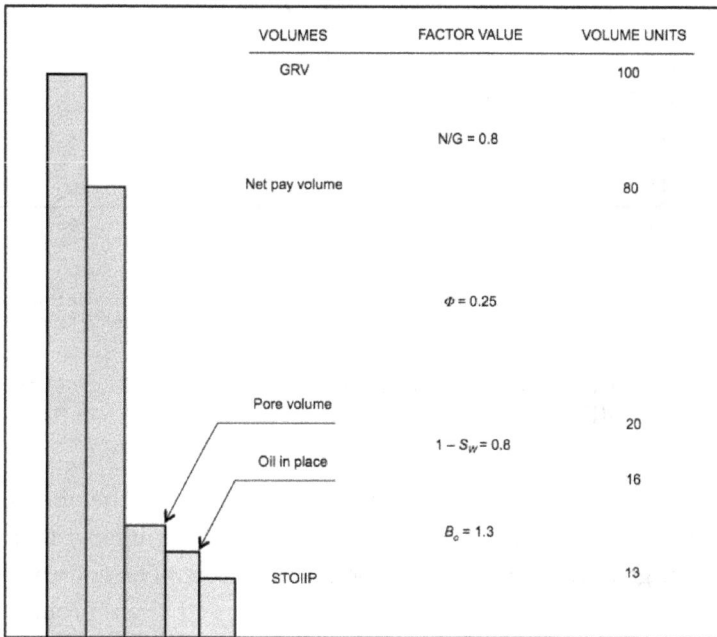

| VOLUMES | FACTOR VALUE | VOLUME UNITS |
|---|---|---|
| GRV | | 100 |
| | N/G = 0.8 | |
| Net pay volume | | 80 |
| | $\Phi = 0.25$ | |
| Pore volume | | 20 |
| | $1 - S_W = 0.8$ | |
| Oil in place | | 16 |
| | $B_o = 1.3$ | |
| STOIIP | | 13 |

Figure 6.8. Illustration of STOIIP. It amounts to a small fraction of GRV.

of the reserves. In the *probabilistic* or *stochastic* approach, each parameter in the STOIIP equation is considered as a variable and a Monte Carlo simulation — a technique that enables the impact of risk and uncertainty to be assessed — is used to produce a range of values for the reserves through numerical operations on random numbers.

In the simulation it is assumed that each variable varies between certain limits. The frequency of the occurrence of a particular value of a given variable together with its minimum and maximum limits and its most likely value define the shape of its distribution, which is either rectangular or triangular, as shown in Fig. 6.9.

A rectangular or uniform distribution means that any numerical value of the variable is equally likely to occur within an upper and a lower limit. A triangular distribution signifies that there is a strong probability for the occurrence of a certain value of the given variable. The shapes of the probability distributions and the minimum, most

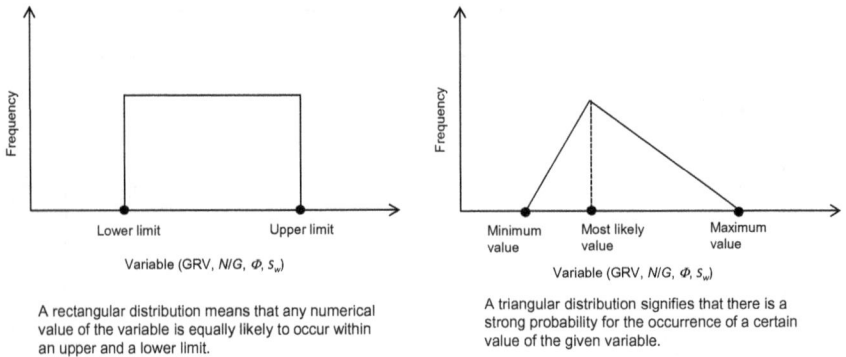

A rectangular distribution means that any numerical value of the variable is equally likely to occur within an upper and a lower limit.

A triangular distribution signifies that there is a strong probability for the occurrence of a certain value of the given variable.

Figure 6.9. Illustrations of rectangular and triangular distributions of the variables in the STOIIP equation.

likely and maximum values of the parameters in the volumetric equation are generally determined in multidisciplinary teams of geologists, geophysicists and petroleum engineers. The software then makes repeated random selections of values and their probabilities from each set of variables, and produces a large number (maybe several thousand) of volumetric estimates.

The rectangular or the triangular frequency distributions are converted into probability distribution and cumulative probability curves from which the minimum (P90), most likely (P50) and maximum (P10) reserves values may be estimated. Examples of probability distribution and cumulative probability curves are shown in Figs. 6.10 and 6.11. In this case, there is a 90% chance that the size of the reserves is about 130 million barrels but only a 10% chance that it is around 480 million barrels. The most likely volume of the reserves is about 310 million barrels. Field development plans and the commercial valuation of the project are based on the P50 estimate of the reserves.

The results of an actual case study are presented in Figs. 6.12 and 6.13. Eight leads were evaluated and the minimum, maximum and most likely values of the parameters used in the iterations are shown in Fig. 6.12. In some cases, mean values of certain parameters were found to be sufficient. The cumulative probability distribution curve relating to the leads is shown in Fig. 6.13. It indicates a

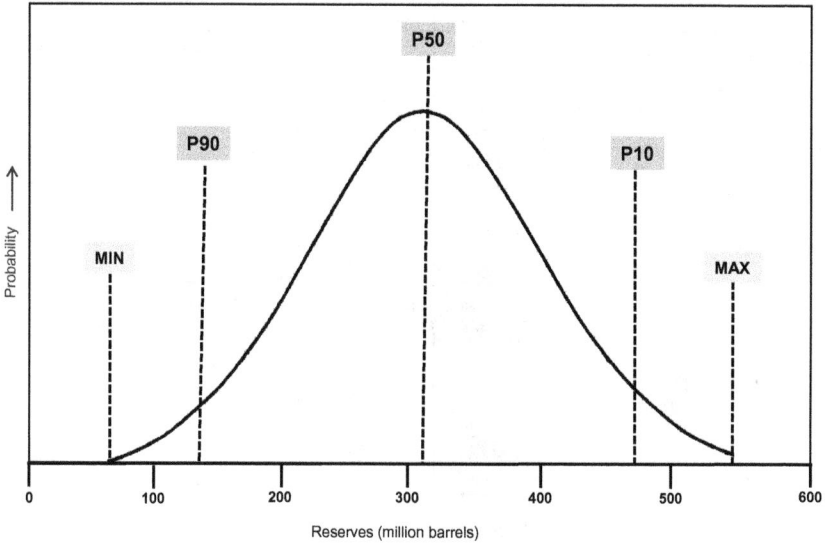

Figure 6.10.  A generic probability distribution curve.

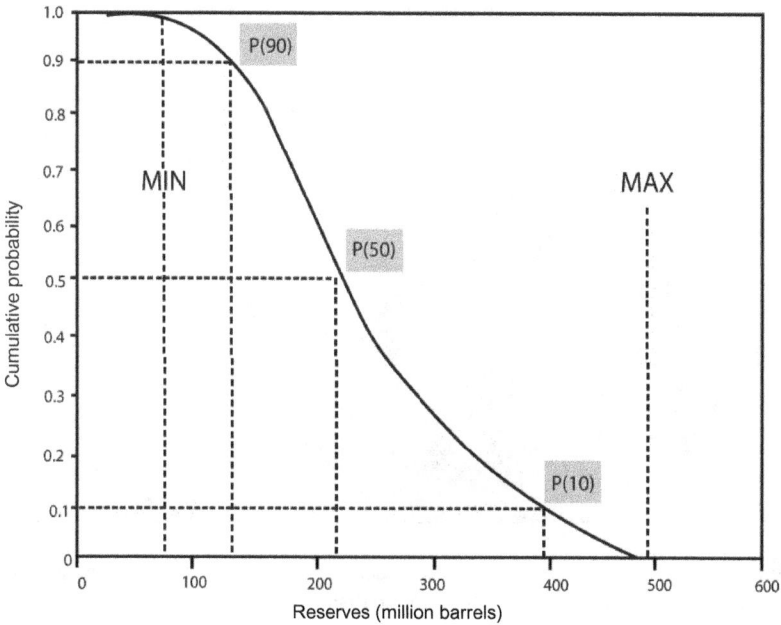

Figure 6.11.  A generic cumulative probability distribution curve.

| | Area | Thickness | Net-to-Gross | Porosity | Sw | Bo | RF |
|---|---|---|---|---|---|---|---|
| Lead B | 0 - 2964 | 10, 50, 60 | 0.5, 0.8, 0.9 | M0.2 | 0.2, 0.3, 0.5 | M1.4 | M0.3 |
| Lead C | M 865 | 10, 50, 60 | 0.5, 0.9, 0.99 | M0.25 | 0.2, 0.3, 0.5 | M1.6 | M0.35 |
| Lead D | M 618 | 10, 50, 60 | 0.5, 0.9, 0.99 | M0.25 | 0.2, 0.3, 0.5 | M1.6 | M0.35 |
| Lead E | M 988 | 10, 50, 60 | 0.5, 0.9, 0.99 | M0.25 | 0.2, 0.3, 0.5 | M1.6 | M0.35 |
| Lead F | M 4446 | 10, 50, 60 | 0.5, 0.9, 0.99 | M0.25 | 0.2, 0.3, 0.5 | M1.6 | M0.35 |
| Lead G | 0 - 1235 | 10, 50, 60 | 0.5, 0.9, 0.9 | M0.25 | 0.2, 0.3, 0.5 | M1.6 | M0.35 |
| Lead H | M 988 | 10, 50, 60 | 0.5, 0.9, 0.99 | M0.25 | 0.2, 0.3, 0.5 | M1.6 | M0.35 |
| Lead I | M 988 | 10, 50, 60 | 0.5, 0.9, 0.99 | M0.25 | 0.2, 0.3, 0.5 | M1.6 | M0.35 |

Figure 6.12. Ranges of the values of the parameters used in evaluating eight leads in Nigeria (courtesy PetroVision).

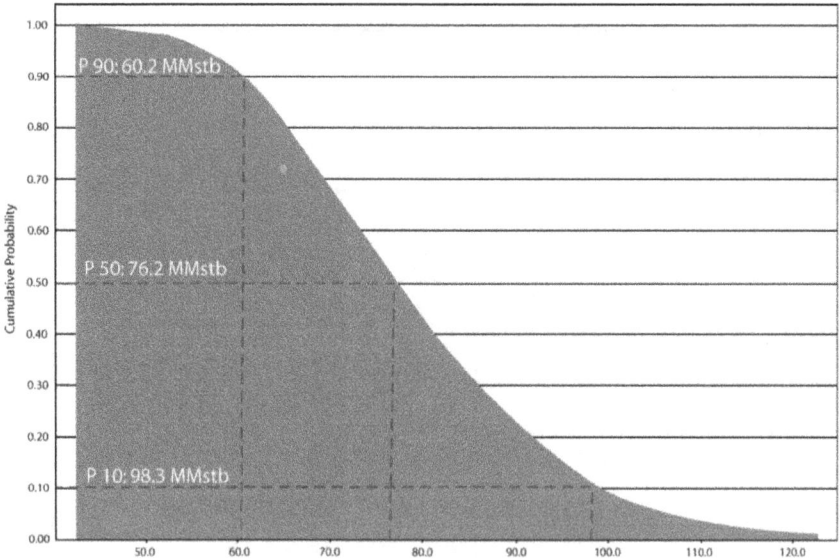

Figure 6.13. A cumulative probability reserves distribution curve relating to the leads detailed in Fig. 6.12 (courtesy PetroVision).

P90 value of 60.2 MMstb, a P10 value of 98.3 MMstb and a most likely reserve of 76.8 MMstb.

## 6.4.   Oil and Gas Reserves, Production and Consumption

### 6.4.1.   *Introduction*

There are several publicly available sources of information on oil and gas reserves, production and consumption, in global terms as well as for individual countries. The most widely used ones include the Oil and Gas Journal, BP's Statistical Review of World Energy, the US Energy Information Administration (EIA), and ENI's World Oil and Gas Review. They use various sources of information and different reporting procedures, and the statistics quoted can vary significantly from one publication to another. Often the information is collected from state oil companies or ministries who respond by filling in questionnaires sent to them annually by those compiling the statistics. Since verification of the data is impossible, the information is published without being independently audited. Also, in recent years, the distinction between conventional and unconventional reserves has become blurred and some countries, notably Venezuela and Canada, include both categories in their recoverable reserve figures. This has propelled Venezuela to the top of the list of the world's reserve holders, displacing Saudi Arabia from that position. Currently, the most recent statistics available relate to late 2015 and a selection of these is presented and briefly discussed below.

### 6.4.2.   *Oil reserves*

The world's total oil reserves and the top ten reserve holders are listed in Table 6.2.

As mentioned in Sec. 6.4.1, with a reported reserve figure of nearly 298 billion barrels, Venezuela tops the list. There has been substantial growth in oil reserves in global terms in the 21st century;

Table 6.2. Total global oil reserves and the top ten reserve holders in millions of barrels, as of 1 January 2016 (ENI World Oil and Gas review, 2016).

| | 2000 | 2005 | 2010 | 2011 | 2012 | 2013 | 2014 | 2015 | Δy/y (2015–2014) | CAGR (2015–2000) |
|---|---|---|---|---|---|---|---|---|---|---|
| Venezuela | 76,848 | 80,012 | 296,501 | 297,571 | 297,735 | 298,350 | 299,953 | 300,878 | 0.3% | 9.5% |
| Saudi Arabia | 262,766 | 264,211 | 264,516 | 265,405 | 265,850 | 265,789 | 266,578 | 266,455 | 0.0% | 0.1% |
| Canada | 181,200 | 178,792 | 175,214 | 173,625 | 173,105 | 173,200 | 172,481 | 170,863 | -0.9% | -0.4% |
| Iran | 99,530 | 136,270 | 151,170 | 154,580 | 157,300 | 157,800 | 157,530 | 158,400 | 0.6% | 3.1% |
| Iraq | 112,500 | 115,000 | 143,100 | 141,350 | 140,300 | 144,211 | 143,069 | 142,503 | -0.4% | 1.6% |
| Kuwait | 96,500 | 101,500 | 101,500 | 101,500 | 101,500 | 101,500 | 101,500 | 101,500 | 0.0% | 0.3% |
| United Arab Emirates | 97,800 | 97,800 | 97,800 | 97,800 | 97,800 | 97,800 | 97,800 | 97,800 | 0.0% | 0.0% |
| Russia | 48,573 | 60,000 | 60,000 | 60,000 | 80,000 | 80,000 | 80,000 | 80,000 | 0.0% | 3.4% |
| Libya | 36,000 | 41,464 | 47,097 | 48,014 | 48,472 | 48,363 | 48,363 | 48,363 | 0.0% | 2.0% |
| United States | 23,517 | 23,019 | 25,181 | 28,950 | 33,403 | 36,520 | 39,933 | 43,629 | 9.3% | 4.2% |
| **The World Top 10** | **1,035,234** | **1,098,068** | **1,362,079** | **1,368,795** | **1,395,465** | **1,403,533** | **1,407,207** | **1,410,391** | **0.2%** | **2.1%** |
| Rest of the World | 204,086 | 219,339 | 249,755 | 250,858 | 255,387 | 256,595 | 256,883 | 257,474 | 0.2% | 1.6% |
| **World** | **1,239,320** | **1,317,407** | **1,611,834** | **1,619,653** | **1,650,852** | **1,660,128** | **1,664,090** | **1,667,865** | **0.2%** | **2.0%** |

| 2000 | 2015 |
|---|---|
| **World: 1,239,320** million barrels as at 31ˢᵗ December | **World: 1,667,865** million barrels as at 31ˢᵗ December |

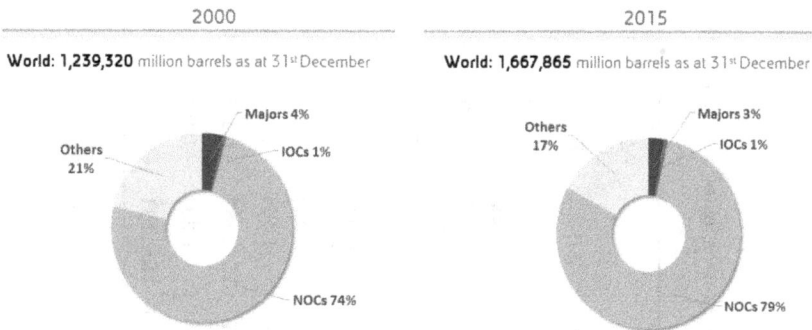

Figure 6.14. Proportions of total global oil reserves controlled by various operators (ENI World Oil and Gas review, 2016).

global reserves grew from just over 1.24 trillion barrels in 2000 to nearly 1.67 trillion barrels as of the end of 2015, an increase of 37%. It should be noted that the top ten reserve holders' share amounted to 85% of the global total. Furthermore, as shown in Fig. 6.14, despite their size and the global spread of their activities, the private, non-state oil companies control only a tiny fraction of the world's reserves, 79% of which was owned by national oil companies (NOCs) in 2015.

### 6.4.3.  *Natural gas reserves*

The total global natural gas reserves and the top ten reserve holders are listed in Table 6.3.

As with oil, the 21ˢᵗ century has witnessed a significant growth in natural gas reserves in global terms. The reserves grew from just under 157.5 trillion cubic metres (TCM) in 2000 to 198.3 TCM as of the end of 2015, an increase of 28%. The top ten reserve holders' share amounted to 78% of the global total. Figure 6.15 shows that, as with oil, the private sector oil companies control only a tiny proportion of the world's reserves, 80% of which was in the hands of NOCs in 2015.

Table 6.3. Total global natural gas reserves and the top ten reserve holders in billions of cubic metres, as of 1 January 2016 (ENI World Oil and Gas review, 2016).

| | 2000 | 2005 | 2010 | 2011 | 2012 | 2013 | 2014 | 2015 | Δy/y (2015–2014) | CAGR (2015–2000) |
|---|---|---|---|---|---|---|---|---|---|---|
| Russia | 43,809 | 44,860 | 46,000 | 48,676 | 48,810 | 49,335 | 49,896 | 50,485 | 1.2% | 1.0% |
| Iran | 26,000 | 27,580 | 33,090 | 33,620 | 33,780 | 34,020 | 34,020 | 33,500 | −1.5% | 1.7% |
| Qatar | 14,443 | 25,636 | 25,201 | 25,110 | 25,069 | 24,681 | 24,531 | 24,299 | −0.9% | 3.5% |
| Turkmenistan | 2,680 | 2,680 | 10,000 | 10,000 | 9,967 | 9,934 | 9,904 | 9,904 | 0.0% | 9.1% |
| United States | 5,021 | 5,784 | 8,621 | 9,454 | 8,717 | 9,573 | 10,434 | 8,630 | −17.3% | 3.7% |
| Saudi Arabia | 6,301 | 6,900 | 8,016 | 8,151 | 8,235 | 8,317 | 8,489 | 8,588 | 1.2% | 2.1% |
| United Arab Emirates | 6,060 | 6,060 | 6,091 | 6,091 | 6,091 | 6,091 | 6,091 | 6,091 | 0.0% | 0.0% |
| Venezuela | 4,152 | 4,315 | 5,525 | 5,528 | 5,563 | 5,581 | 5,617 | 5,702 | 1.5% | 2.1% |
| Nigeria | 4,106 | 5,154 | 5,110 | 5,176 | 5,118 | 5,107 | 5,324 | 5,284 | −0.7% | 1.7% |
| Algeria | 4,523 | 4,504 | 4,504 | 4,504 | 4,504 | 4,504 | 4,504 | 4,504 | 0.0% | 0.0% |
| **The World Top 10** | **117,095** | **133,473** | **152,158** | **156,310** | **155,055** | **157,143** | **158,811** | **156,987** | **−1.1%** | **2.0%** |
| Rest of the World | 40,447 | 40,904 | 42,357 | 41,900 | 41,929 | 41,995 | 41,661 | 41,294 | −0.9% | 0.1% |
| **World** | **157,542** | **174,377** | **194,515** | **198,210** | **197,784** | **199,138** | **200,471** | **198,281** | **−1.1%** | **1.5%** |

| 2000 | 2015 |
|------|------|
| **World: 157,542** billion cubic metres as at 31ˢᵗ December | **World: 198,281** billion cubic metres as at 31ˢᵗ December |

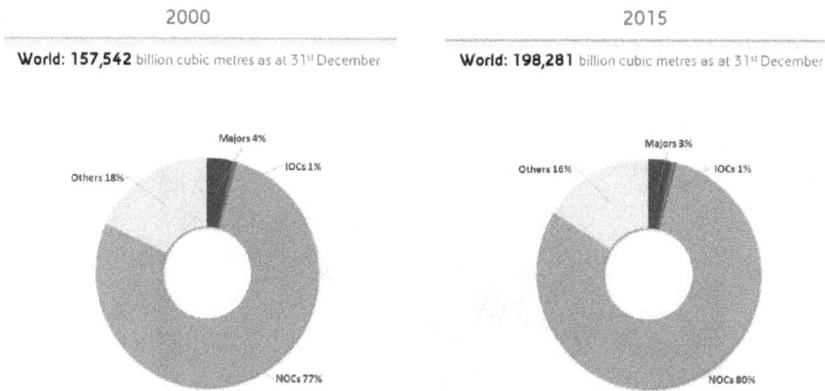

Figure 6.15. Proportions of total global natural gas reserves controlled by various operators (ENI World Oil and Gas review, 2016).

### 6.4.4.  *Global crude oil production and consumption*

Global crude oil production from 1989 to the end of 2014 is shown in Fig. 6.16. World oil output reached about 89 million barrels per day (b/d) in 2014, an increase of 2.1 million b/d compared to 2013, all of which was in non-OPEC countries. US output grew by 1.6 million b/d, its largest increase on record. This is due to supplies of shale oil and will continue to increase over the next decade. Nonetheless, as depicted in Fig. 6.17, 60% of total global oil was produced by NOCs in 2015.

Figure 6.18 presents the worldwide oil consumption trend for the period 1989–2014. In 2014, global consumption was around 92 million b/d, an increase of 840,000 b/d compared to 2013, with the emerging economies accounting for all of the net increase.

### 6.4.5.  *Global natural gas production and consumption*

Worldwide natural gas production during the period 1989–2014 is shown in Fig. 6.19. Global natural gas output reached 3,461 billion cubic metres (BCM) in 2014, an increase of 1.6% compared to 2013. With an increase of 6.1%, the US remained the world's leading producer. In the case of gas, the global production shares of the NOCs and non-state sector each amounted to 49% in 2015 (Fig. 6.20).

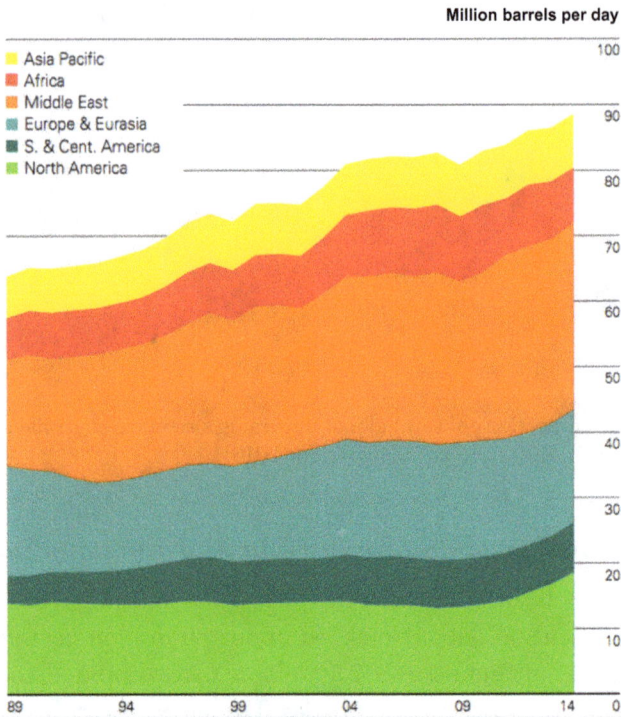

Figure 6.16. Global crude oil production, 1989–2014 (BP Statistical Review of World Energy, 2015).

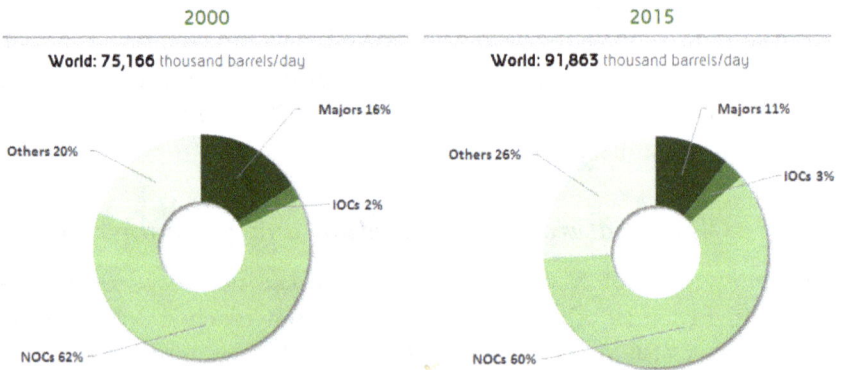

Figure 6.17. Proportions of global oil produced by various operators (ENI World Oil and Gas review, 2016).

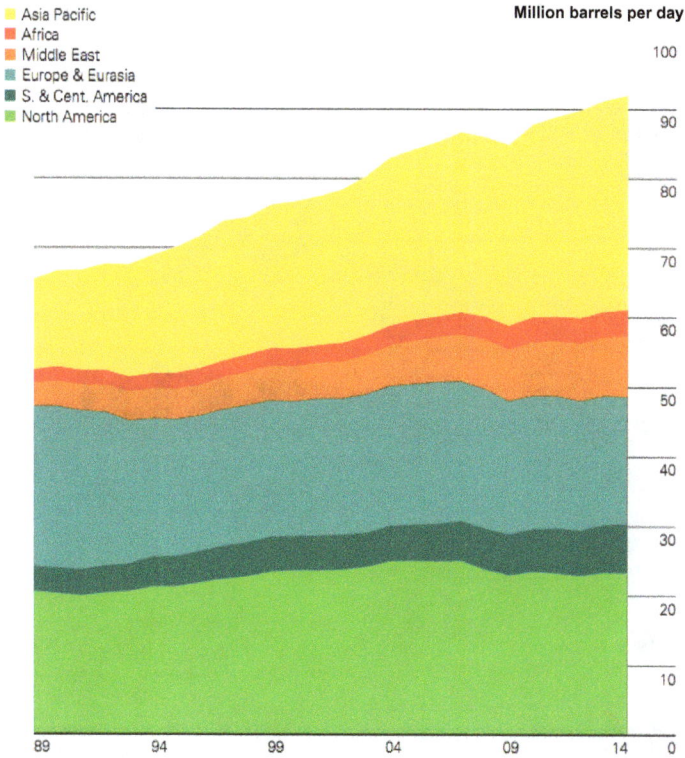

Figure 6.18. Global crude oil consumption, 1989–2014 (BP Statistical Review of World Energy, 2015).

The global natural gas consumption trend for the period 1989–2014 is shown in Fig. 6.21. In 2014, global gas consumption amounted to 3,393.0 BCM, an increase of 0.4% over 2013, with the US and China recording the largest growth.

### 6.4.6.  *Biofuels*

Biofuels include ethanol (known popularly as alcohol or spirits), petrol and diesel extracted from plants, and their use as transport fuel is being promoted in Europe and North and South America. Globally, the US and Brazil dominate biofuel production, mostly in the form of ethanol, as shown in Fig. 6.22.

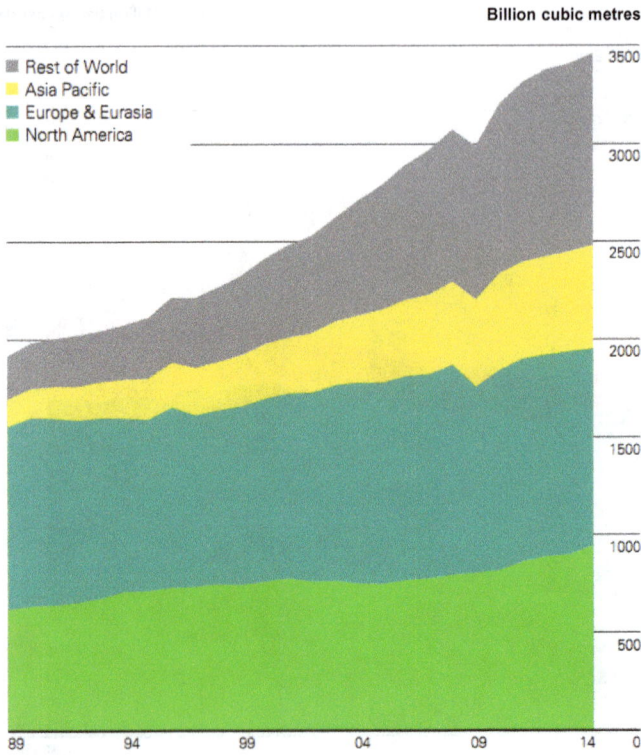

Figure 6.19. Global natural gas production, 1989–2014 (BP Statistical Review of World Energy, 2015).

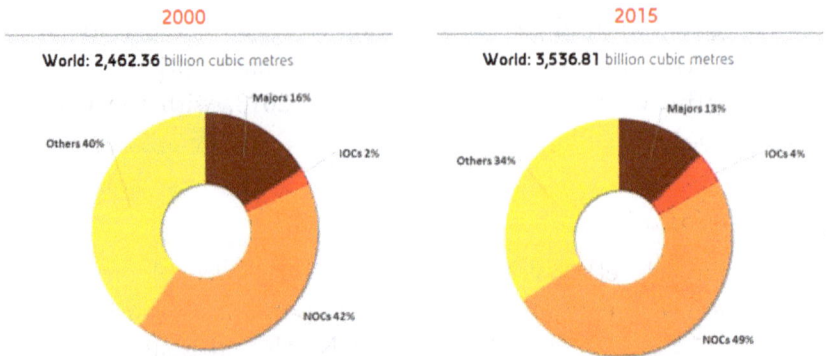

Figure 6.20. Proportions of global natural gas produced by various operators (ENI World Oil and Gas review, 2016).

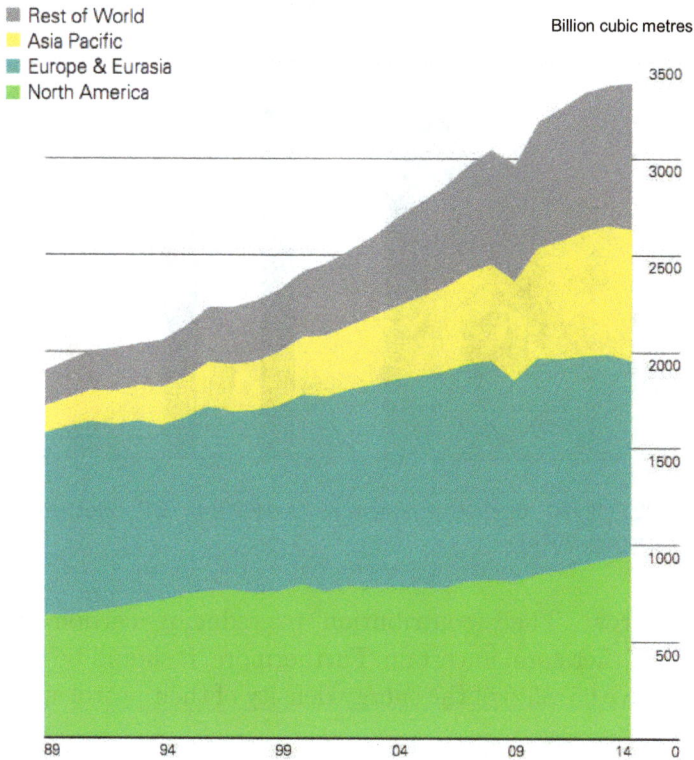

Figure 6.21. Global natural gas consumption, 1989–2014 (BP Statistical Review of World Energy, 2015).

World biofuels production grew by 7.4% in 2014 compared to 2013. Global ethanol output increased by 6.0%, the second consecutive year of growth, led by increases from North America, South and Central America and Asia Pacific. Biodiesel production increased by 10.3% in 2014. Despite this growth, biofuels accounted for only 1% of the world's energy consumption in 2014.

Presenting biofuels as environmentally friendly alternatives to fossil fuels is, however, open to debate. Large-scale deforestation has taken place in a significant number of countries (Brazil, Indonesia and Paraguay, to mention a few) to supply the biofuel market. Some studies indicate that trees and forest soils are more effective at trapping carbon from the atmosphere than the biofuel plants

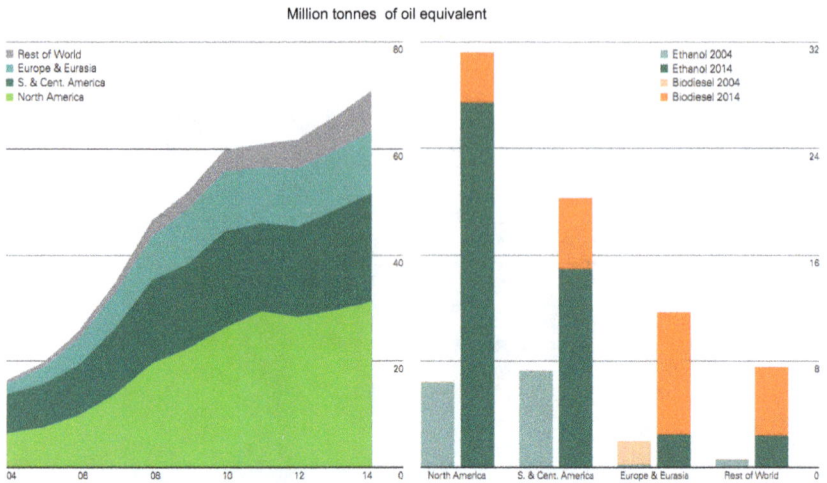

Figure 6.22. Global biofuel production, 2004–2014 (BP Statistical Review of World Energy, 2015).

as they grow. Their contribution to reducing carbon emissions globally is, therefore, uncertain. Furthermore, it should be noted that biofuels have 65–85% of the energy density of their petroleum-derived counterparts.

## References

BP Statistical Review of World Energy, 2015 [online]. Available at: bp.com/statisticalreview.

Dikkers, A.J. (1985). *Geology in Petroleum Production*, Elsevier Scientific Publishing Company, London, UK.

ENI World Oil and Gas review, 2016 [online]. Available at: eni.com/world oil-gas-review-2016.

Selley, R.C. (1997). *Elements of Petroleum Geology*, 2nd Edition, Academic Press, Waltham, MA.

SPE-PRMS (2011). *Guidelines for Application of the Petroleum Resources Management System* [online]. Available at: http://www.spe.org/indus try/docs/PRMS_Guidelines_Nov2011.pdf.

Stoneley, R. (1995). *An Introduction to Petroleum Geology for Non-Geologists*, Oxford University Press, Oxford, UK.

White, D.A. and Gehman, H.M. (1979). Methods of estimating oil and gas resources, *AAPG Bull.*, **63**, 2183–2192.

# Chapter 7

# The Unconventionals:
# Oil Shale, Shale Oil, Shale Gas, Oil Sands,
# Coal Bed Methane and Gas Hydrates

## 7.1. Introduction

The term *unconventional hydrocarbons* refers to oil and gas deposits
that are not stored in pore spaces in permeable rocks (the reservoir)
and are not commercially recoverable by "conventional" drilling
and production methods. According to this definition, oil shale,
shale oil, shale gas, oil sands, coal bed methane and gas hydrates
fall into the unconventional category. However, as mentioned in
Chapter 6, advances in drilling and extraction technology since the
mid-1990s have made it possible for some of these hydrocarbons to
be produced economically, resulting in their status being changed
from unconventional to conventional. Mention was made in Chapter
6 of the SPE-PRMS document (2011), establishing guidelines and
procedures to standardise the assessment, classification and reporting
of oil and gas resources. The relationship between the conventional
and unconventional resources is illustrated by a 'resource triangle',
as depicted in Fig. 7.1. Oil sands, heavy oil and tight gas formations
straddle the boundary. Despite the enormous volumes of methane
associated with gas hydrates (see below), exploitation of this resource
is not yet technically and economically feasible. These unconventional
hydrocarbons are considered briefly below.

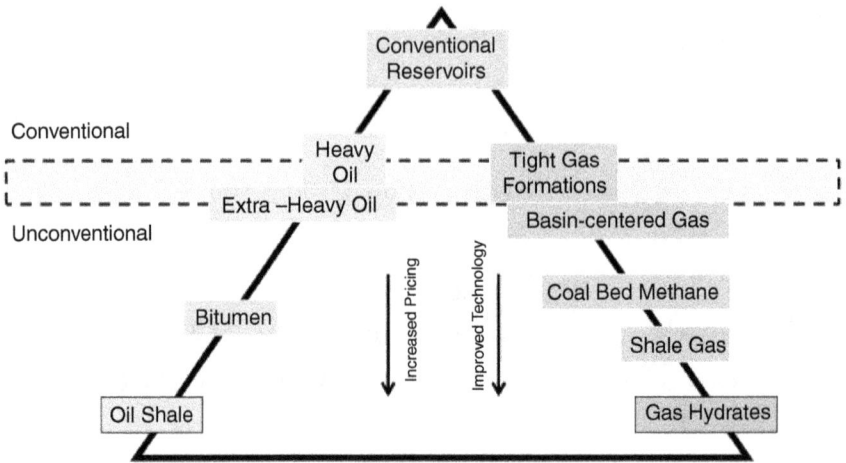

Figure 7.1. Resource triangle, illustrating the relationship between conventional and unconventional resources (from SPE-PRMS, 2011).

## 7.2.   Oil Shale, Shale Oil and Shale Gas

That vast amounts of oil and gas are locked up in dark-coloured, organic-rich shale rocks has been known for nearly two centuries. The earliest record of gas production from fractured shale dates from 1821 in western New York State, close to the shore of Lake Erie, and rapidly expanded to other areas in the Appalachian region. Commercial production of oil from shale began in 1838 in France. The oil was extracted by mining and heating the shale in a retort and marked the dawn of the oil shale industry. This was followed by the launch of the Scottish oil shale industry in 1859, in a region immediately to the west and southwest of Edinburgh, covering an area of 50 sq. mi. The Scottish oil shale industry survived for more than 100 years. At its peak in 1913, it employed 10,000 people and until 1962 the oil produced was exempt from excise duty. The withdrawal of this concession and the availability of abundant supplies of "cheap oil" made the operation uneconomic and the enterprise closed down in 1964.

Oil- and gas-producing shales are mature source rocks in the oil and gas windows. They have generated and released hydrocarbons to nearby reservoirs, forming conventional accumulations. However, as discussed in Chapter 4, hydrocarbon expulsion from source rocks is an inefficient process and much of the oil and gas generated remains in the source rocks. Liberating these hydrocarbons from the source rock economically had eluded the industry until the advent of horizontal drilling and fracking. This technique has propelled shale oil and shale gas into prominence from being previously neglected sources of these hydrocarbons. Since the oil and gas occur in other low-permeability rocks as well as shale, they are referred to by some investigators as "tight oil" or "tight gas" rather than shale oil or shale gas. It should be noted that only mature source rocks are capable of directly producing hydrocarbons. Because they have not generated any hydrocarbons, immature source rocks have no direct oil and gas production potential. They will yield hydrocarbons only on being subjected to heat to replicate the natural process. Oil shales are immature source rocks.

Due to the low permeability of the shale, it requires fractures to allow oil and gas to flow at commercial rates. This is accomplished by drilling long horizontal wells into shale layers and fracturing them by pumping water containing additives under very high pressures. On average, some 20 million L of water are forced under pressure into each well, combined with large volumes of sand to help keep the fractures open, plus 200,000 L of chemical additives. The process, which is therefore very water-intensive, is illustrated diagrammatically in Fig. 7.2.

Tight oil and gas production has risen rapidly in the US since the turn of the present century and shale gas is projected to account for up to half of the natural gas output in North America by 2020. The US is now self-sufficient in gas and its oil production has risen substantially, resulting in a significant reduction in its imports of crude oil.

A characteristic feature of tight oil and gas wells is their high rates of decline. The intrinsically low permeability of the formations means that each well drains only a limited volume of the rock and this accounts for the steep decline in the rate of production. A good oil well may produce up to 700 b/d initially, but that may drop to 100 b/d by the end of the first year and the production remains

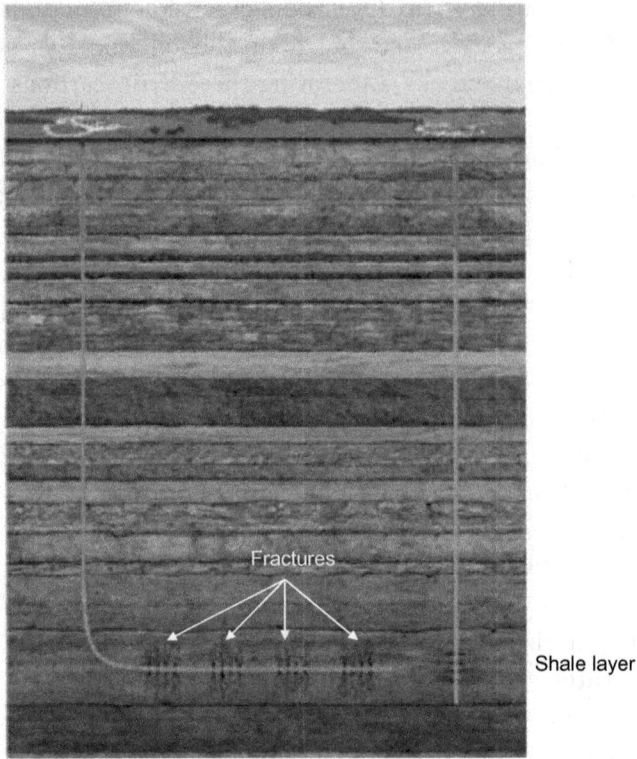

Figure 7.2. Production of oil and gas from shale (http://energy.gov/fe/shale-gas-101).

at this level for the next two or three decades. Consequently, new wells constantly need to be drilled in order to maintain production and a sustained high oil price is therefore required for the economic viability of the operations. This leads to the conclusion that the current rapid growth in US tight oil and gas output cannot be sustained indefinitely. Nonetheless, the advent of plentiful supplies of tight oil and gas has transformed the American energy landscape and its beneficial effects will continue for some time to come.

### 7.2.1.  *Distribution of tight oil and gas resources*

Although tight oil and gas production is a North American phenomenon, hydrocarbon-bearing shale rocks are present in many parts

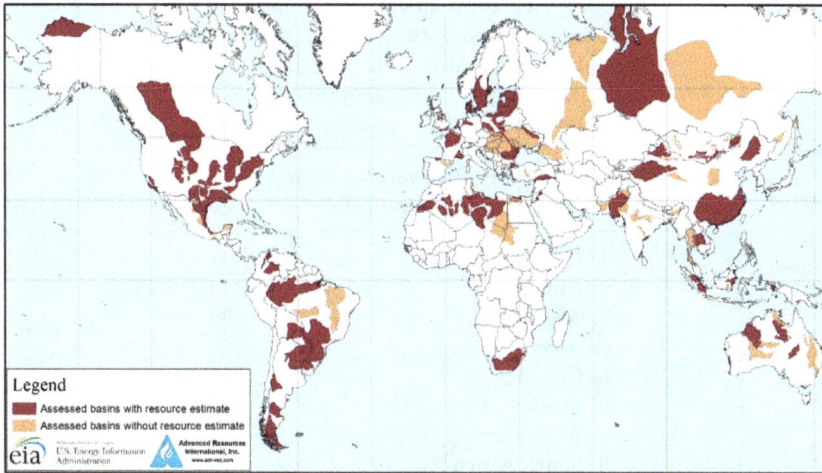

Figure 7.3. Global distribution of shale oil and shale gas formations, as of May 2013 (http://www.eia.gov/analysis/studies/worldshalegas/pdf/overview.pdf).

of the world. Figure 7.3 shows the result of a study in 2013 by the US Energy Information Administration (EIA) of the shale gas resources of 42 countries and Table 7.1 lists the top 10 resource holders as well as the grand total. In terms of shale gas, the US tops the list and the total resources of the 42 countries amount to 7,795 trillion cu ft (TCF). Many countries with large shale gas prospects were not covered by the study and the figure of 7,795 TCF represents only a fraction of the global total amount that could be potentially available. In the EIA list the UK has been included in the "others" category, while an assessment by the British Geological Survey published in December 2014 has placed the in-place gas resources of the UK at between 871 TCF and 2,146 TCF, with a central estimate of 1,420 TCF. It should be emphasised that the estimates vary greatly from source to source depending on the method of assessment.

There is a vast amount of publicly available information on the American tight gas basins and plays, the distribution of which is illustrated in Fig. 7.4.

Table 7.2 presents the result of the 2013 EIA study relating to the shale oil resources of the same 42 countries. In the case of oil, Russia

Table 7.1. Total shale gas resources of 42 countries and the top ten resource holders as of May 2013 (http://www. eia.gov/analysis/studies/worldshalegas/ pdf/overview.pdf).

| | Technically Recoverable Shale Gas Resources (Tcf) |
|---|---|
| 1. US | 1,161 |
| 2. China | 1,115 |
| 3. Argentina | 802 |
| 4. Algeria | 707 |
| 5. Canada | 573 |
| 6. Mexico | 545 |
| 7. Australia | 437 |
| 8. South Africa | 390 |
| 9. Russia | 285 |
| 10. Brazil | 245 |
| 11. Others | 1,535 |
| **TOTAL** | **7,795** |

tops the list and the total resources of the 42 countries amount to 335 billion barrels. Again, since many countries with significant shale oil prospects were excluded, the figure of 335 billion barrels represents only a fraction of the global total amount that could be potentially available.

## 7.3.   Oil or Tar Sands

The largest known concentrations of oil or tar sands are in Canada and Venezuela, the magnitude of the resources in the two countries being estimated at up to 3.5 trillion barrels of oil in place.

The Canadian deposits, consisting of sand and clay heavily impregnated with highly viscous bitumen, are situated in the northern part of the province of Alberta, centred largely on an area known as Athabasca, the location of which is shown in Fig. 7.5. In the early stages of the operation, the sand was excavated and transported

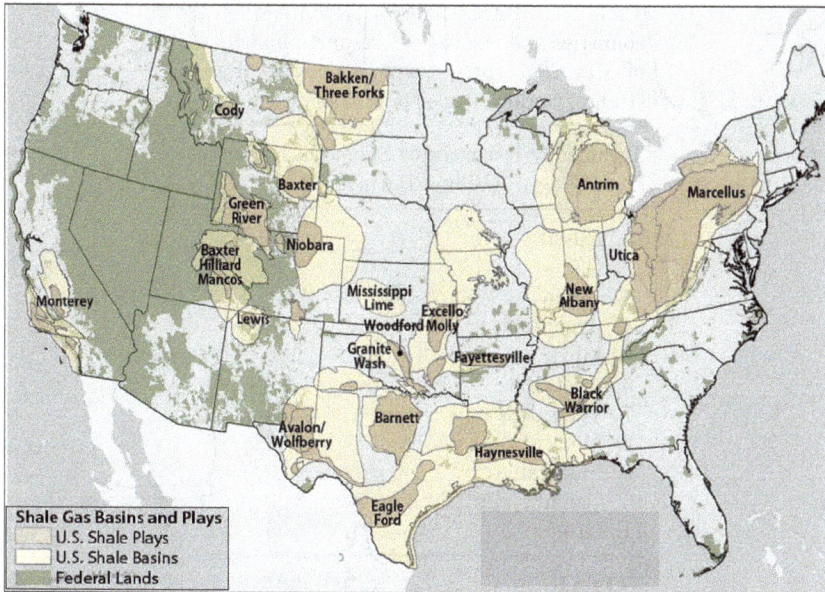

Figure 7.4. Distribution of the US gas basins and plays as of May 2013 (http://www.eia.gov/pub/oil_gas/natural_gas/analysis_publications/maps/maps.htm).

on fixed conveyer belts to processing plants where the bitumen was separated. From the late 1990s, the mining process was modernised and expanded in scale, and giant trucks were used to convey the sand to the separators. The bitumen was then upgraded into a high-quality, low-sulphur "synthetic crude" or "syncrude", which can be processed in a conventional refinery into petrol, diesel, aviation fuel and other products. A more recent technological breakthrough has made it possible to produce the bitumen *in situ*, which avoids the environmentally undesirable mining method. *In-situ* production involves injecting super-hot steam, generated by burning natural gas, to heat the bitumen underground. This reduces its viscosity, enabling it to flow via production wells to the surface.

Estimates of the magnitude of the Canadian oil sand resources vary, but the Canadian Association of Petroleum Geologists places it at around 1.7 trillion barrels of oil in place, of which 10% or 170

Table 7.2. Total shale oil resources of 42 countries and the top ten resource holders as of May 2013 (http://www.eia.gov/analysis/ studies/worldshalegas/pdf/ overview.pdf).

Technically Recoverable Shale Oil Resources
(Billion Barrels)

| | |
|---|---|
| 1. Russia | 75 |
| 2. US | 48 |
| 3. China | 32 |
| 4. Argentina | 27 |
| 5. Libya | 26 |
| 6. Australia | 18 |
| 7. Venezuela | 13 |
| 8. Mexico | 13 |
| 9. Pakistan | 9 |
| 10. Canada | 9 |
| 11. Others | 65 |
| **TOTAL** | **335** |

billion barrels are considered recoverable. Production is currently running at about 1.8 m b/d and is expected to double by 2020.

Venezuela's oil resources occur in a region called the Orinoco Oil Belt and cover an area of 54,000 sq. mi (Fig. 7.6). Like the Canadian deposits, they consist of bitumen-impregnated sand and clay. The first attempt to extract oil from these was in the 1970s but the operation was abandoned due to constraints on costs and technology. Technological advances and large-scale investment by western oil companies in the 1990s led to the commencement of production which reached 600,000 b/d within a decade. However, the enterprise was nationalised in 2007 and the Venezuelan government announced the objective of raising production to 2.1 million b/d by 2019. Whether this will be achieved remains to be seen.

The Orinoco Oil Belt is the site of the world's largest oil resources. Oil in place is estimated at 2.5 trillion barrels, of which 267 billion barrels are recoverable. Production is currently running at 600,000 b/d.

Figure 7.5. Location of the Canadian oil sands (www.mining-technology.com).

## 7.4. Coal Bed Methane

Coal is derived from plant material that once grew in swamps, which are stagnant, oxygen-poor bodies of water. Subsidence causes the vegetation to be buried and the oxygen-poor environment ensures

Figure 7.6. Location of the Orinoco Oil Belt, Venezuela (www.economist.com/node/14382525).

its preservation. Compaction and partial decay of the vegetation transforms it into *peat*, which consists of 50% carbon and is used as fuel in many parts of the world. Continued burial is accompanied by increase in pressure and temperature, which results in the transformation of peat into coal.

The process of coal formation in the sub-surface produces large quantities of methane, some of which escapes into the atmosphere over time but vast amounts remain trapped in the coal seams. Producing methane from coal involves drilling rather than mining. As illustrated in Fig. 7.7, wells are drilled into coal beds and the groundwater is pumped out, resulting in a drop in pressure in the vicinity of the wellbore. This causes release of the gas, which flows into the annulus between the casing and the tubing and rises to the surface where it is fed into pipelines. Produced water is either re-injected into a deeper formation or treated and disposed of at the surface.

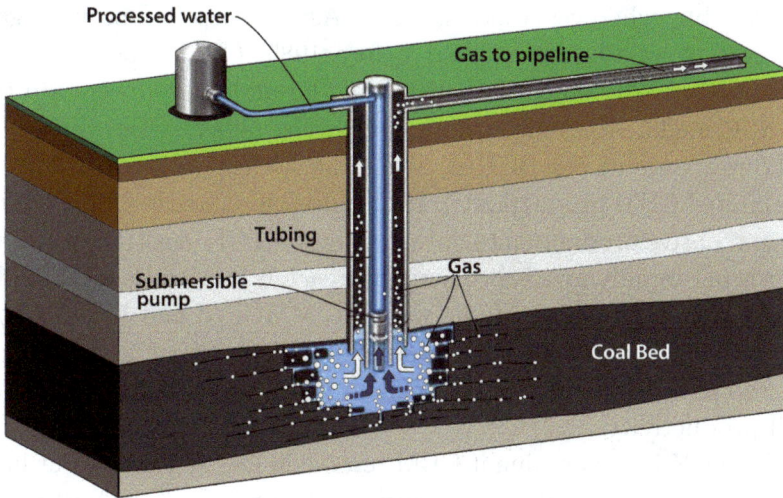

Figure 7.7. Diagrammatic illustration of coal bed methane production.

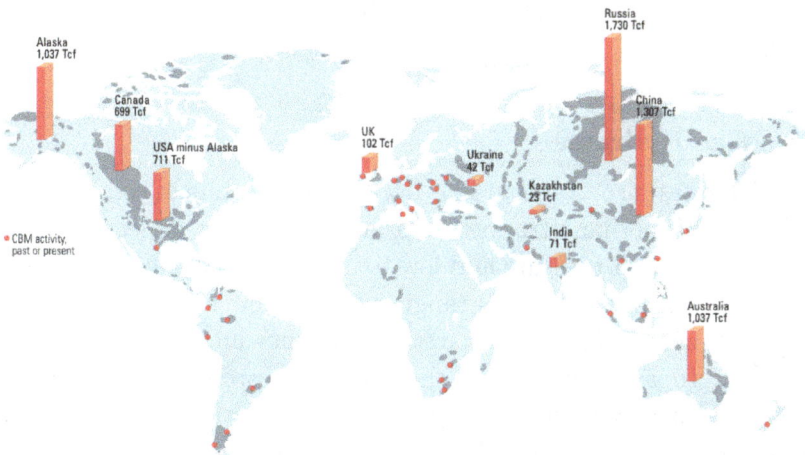

Figure 7.8. Global distribution of coal bed methane reserves (copyright Schlumberger, after Al-Jubori *et al.*, 2009, used with permission).

### 7.4.1. *Global distribution of coal bed methane reserves and production*

Figure 7.8 shows the worldwide distribution of coal bed methane (CBM) reserves. It is an abundant source of natural gas and is available wherever there is coal. As a source of natural gas, however, it was barely utilised before 1985. CBM production has expanded rapidly since the 1990s and the major producers are the USA, Canada, Australia, China, and India. Global CBM output in 2012 was around 5.8 BCF/d. With the largest reserves, the USA dominated CBM production: in 2012 it produced nearly 5 BCF/d and its cumulative production to date is about 20 TCF. Australian output is growing rapidly and is projected to reach 6 BCF/d by 2020. CBM production in China and India was 150 [a]MMCF/d and 10 MMCF/d respectively in 2012. The Chinese and Indian output is far behind the top producers due to more difficult geological conditions and low well productivity.

The UK is the dominant CBM reserve holder in Europe but high costs and the availability of conventional North Sea gas have so far inhibited the development of its CBM resources. Despite the size of its resource base, Russia is yet to establish CBM production. In Southeast Asia, Indonesia has recently announced CBM resources of 453 TCF.

## 7.5.  Gas Hydrates

*Gas hydrates* are a naturally occurring substance composed of molecules of methane trapped in a lattice of water molecules. They are stable at high pressure and low temperature, physically resemble ice and are found at depths ranging from a few hundred to several thousand metres in shallow arctic sediments and in deep oceanic deposits. To a lesser extent, they are also present in permafrost sediments in arctic areas. They occur abundantly worldwide, have been recovered in core samples acquired during the Deep Sea Drilling Project — a study of the world's oceans undertaken during a 20 year period from the mid-1960s until the mid-1980s — and are identifiable on seismic sections and wire line well logs. The global distribution of gas hydrates is illustrated in Fig. 7.9 and a sample of gas hydrate being removed from a core barrel is shown in Fig. 7.10.

---

[a]Each M represents 1,000 CF.

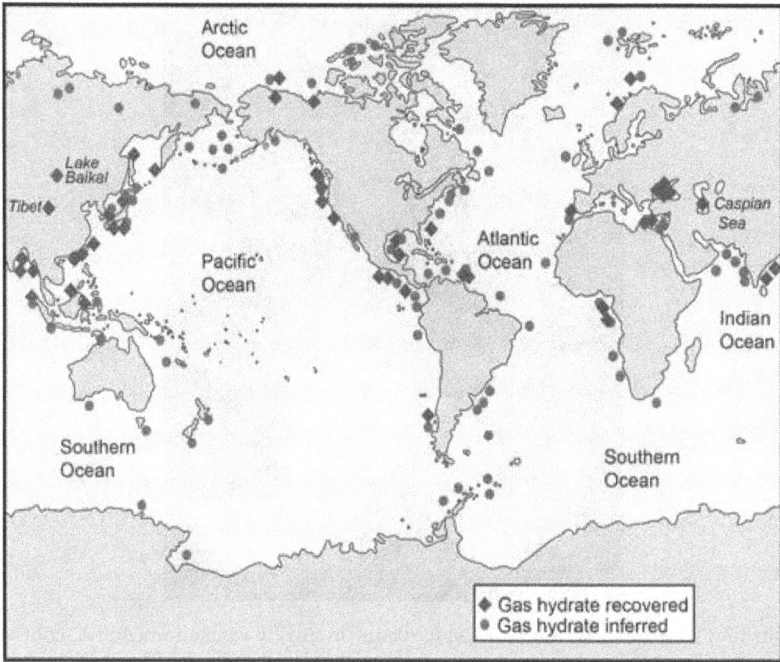

Figure 7.9. Global distribution of gas hydrates (http://soundwaves.usgs.gov).

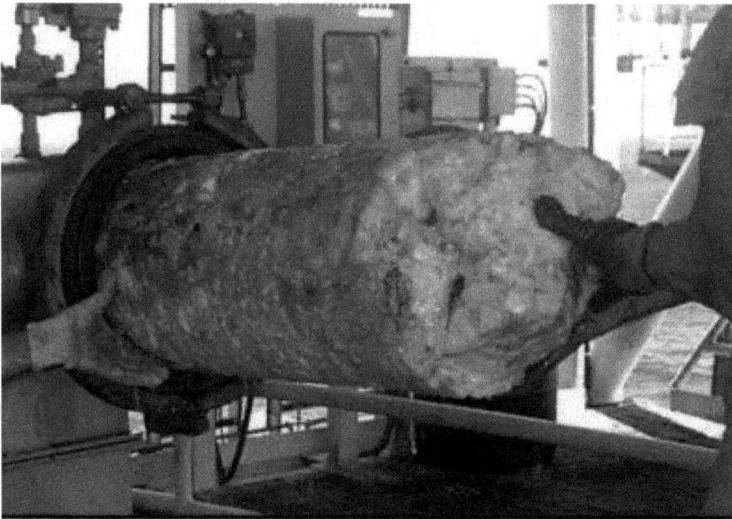

Figure 7.10. A sample of gas hydrate being removed from a core barrel (http://www.chem.ntnu.no).

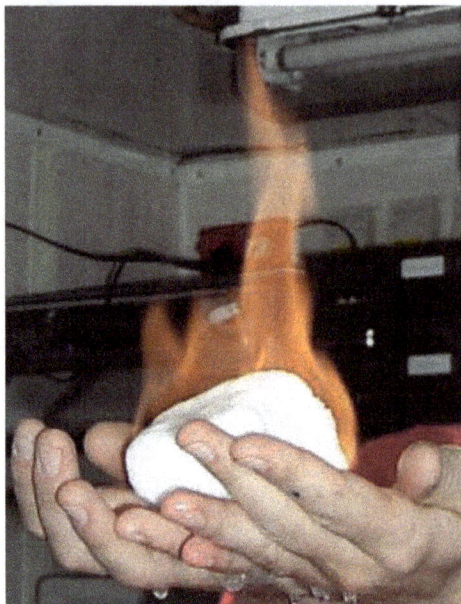

Figure 7.11. A burning lump of gas hydrate (http://www.chem.ntnu.no; courtesy Department of Energy).

The methane in gas hydrate is highly concentrated; 1 cu ft of hydrate contains 164 cu ft of methane gas. Gas hydrate is therefore highly inflammable and burns readily when ignited. A burning sample of gas hydrate is shown in Fig. 7.11.

It is generally accepted that immense quantities of methane are associated with gas hydrates. Indeed, as shown in Fig. 7.12, there may be more organic carbon stored in gas hydrates than in all other reservoired hydrocarbons combined! Prior to 1995, global methane reserves attributed to gas hydrates were between $10^5$ TCF and $10^8$ TCF. However, research carried out during the past two decades by the US Geological Survey has indicated that the amount of gas hydrates in marine sediments is lower than previously thought. This has led to downward revisions of global and regional estimates and most studies now suggest that between $10^5$ TCF and $5 \times 10^6$ TCF of methane is trapped in global gas hydrate deposits. Exploitation of hydrates as a source of natural gas is currently beyond the technical capabilities of the oil industry.

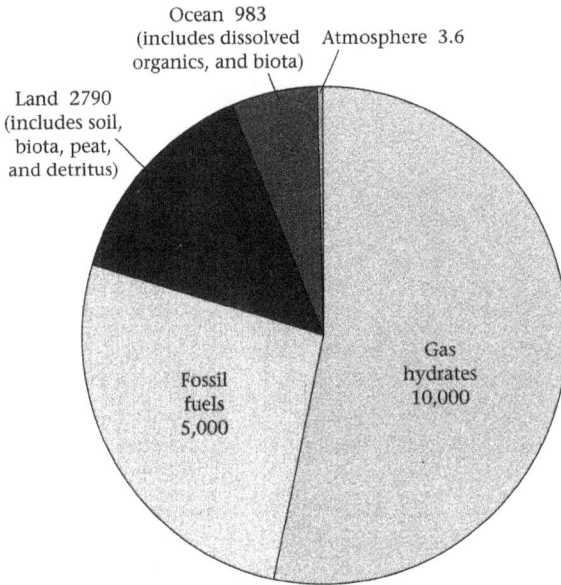

Figure 7.12. Content of organic carbon (in $10^5$ tons) in various materials (from Marshak, 2005).

## References

Al-Jubori, A., Johnson, S., Lambert, S.W. *et al.* (2009). Coalbed methane: Clean energy for the world, *Oilfield Rev.*, **21**(2), 4–13.

Energy Information Administration (2013a). Global shale oil and shale gas distribution map. Available at: http://www.eia.gov/analysis/stud ies/worldshalegas/pdf/overview.pdf.

Energy Information Administration (2013b). US gas basins and plays distribution map. Available at: http://www.eia.gov/pub/oil_gas/natural_gas/analysis_publications/maps/maps.ht.

Energy Information Administration (2013c). Total shale gas resources of 42 countries and the top ten resource holders. Available at: http://www.eia.gov/analysis/studies/worldshalegas/pdf/overview.pdf.

Energy Information Administration (2013d). Total shale oil resources of 42 countries and the top ten resource holders. Available at: http://www.eia.gov/analysis/studies/worldshalegas/pdf/overview.pdf.

Marshak, S. (2005). *Earth: Portrait of a Planet*, 2nd Edition, W.W. Norton, New York, NY.

SPE-PRMS (2011). Guidelines for Application of the Petroleum Resources Management System [online]. Available at: http://www.spe.org/indus try/docs/PRMS_Guidelines_Nov2011.pdf.

# Chapter 8

# Introduction to Open-Hole Logs

## 8.1. Introduction

Well logging techniques were developed in the 1920s and their applications to the study of the rock and fluid properties of the successions penetrated in boreholes revolutionised the exploration and production branches of the petroleum industry. From modest beginnings, the acquisition and interpretation methods of log data advanced rapidly and have now become a complex and sophisticated science and remain an area of active research and development. In addition to the traditional "wireline" techniques, which involve lowering into the well an instrument attached to a cable and measuring the rock and formation fluid properties as the tool is pulled up, the introduction of new devices in the 1980s has made it possible to record these characteristics continuously during drilling. This mode of data acquisition is referred to as "logging while drilling", abbreviated to LWD, and saves time by avoiding the stopping of the drilling process and pulling the drilling equipment out of the well, which is necessary for the running of wireline logs. In terms of quality, LWD data nowadays compare reasonably well with their wireline counterparts.

Until the 1980s, there were a significant number of companies offering logging services at both the national and international levels. However, consolidations in the past three decades have resulted in the disappearance of many well-known names and the industry is

Figure 8.1. Casing.

now dominated by three giant corporations, namely, Schlumberger, Halliburton and Baker Hughes.

Before proceeding to describe log data acquisition and interpretation, it should be noted that oil and gas wells are usually cased. Casing consists of strings of steel pipe (Fig. 8.1), manufactured to standard specifications and cemented into place. As illustrated in Fig. 8.2, each string is cemented all the way to the surface with the diameter of the casing diminishing with depth. It is important that the casing is centralised in the borehole prior to cementing and this is achieved by fastening *centralisers* (Fig. 8.2) to the outside of each string of pipe. Casing generally protects the borehole wall, prevents it from collapse in unconsolidated formations, and shuts off the flow of unwanted fluids into the well. Hydrocarbons are produced through perforations in the production section of the string.

## 8.2. Definition and Types of Logs

A *log* is defined as a continuous recording versus depth of a set of curves representing various properties of the formations penetrated

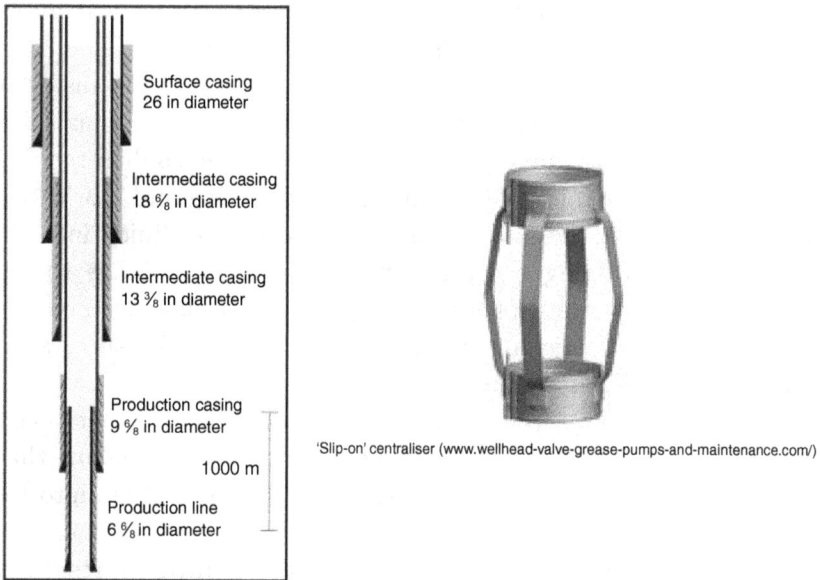

Figure 8.2. Profile of a typical cased borehole and a 'slip-on' centraliser.

in wells. Broadly, they fall into two categories: (a) open-hole and (b) cased-hole types. Open-hole logs are recorded before any casing is run and provide information on the rock and fluid properties, i.e. the lithology, porosity and pore fluid characteristics of the successions encountered in wells. They are used by both petroleum geoscientists and engineers.

Cased-hole logs, also referred to as *production logs*, are run in cased holes and measure flow rates in production wells. The logs also monitor the quality of the cement bonding the casing to the borehole wall and are used primarily by petroleum engineers.

This discussion is concerned with the acquisition and certain aspects of the interpretation of the open-hole logs only; cased-hole logs are beyond the scope of its coverage.

### 8.2.1. *Invasion effects*

Before proceeding further, however, mention must be made of the consequences of the invasion of permeable beds (i.e. potential reservoirs) by the fluid that is used in drilling, which is a major consideration in log data acquisition and interpretation. The fluid, usually a water- or oil-based mud, is pumped down the inside of the drill pipe and returns to the surface through the annular space between the drill pipe and the sides of the borehole, as illustrated in Fig. 8.3. It has several functions that include cooling and lubricating the bit, preventing the flow of unwanted formation fluids into the borehole and bringing the cuttings produced by drilling to the surface. Figure 8.4 presents a photograph of cuttings, which are used by the wellsite geologist to identify the lithology of the successions being penetrated in the borehole.

Regardless of its type, the mud will always invade a permeable formation, since it is kept in the well bore under greater pressure than that of the fluid in the formation pores. As the mud filters into the porous layers, some of the material suspended in the mud becomes deposited as a *mudcake* on the porous rock faces. Invasion results in

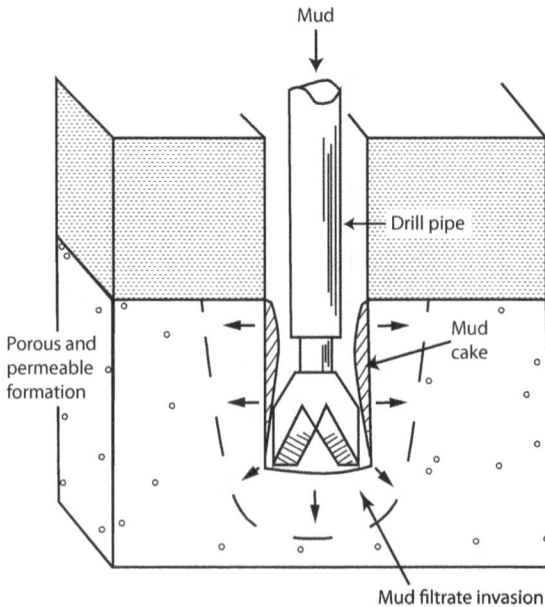

Figure 8.3. Effect of invasion into a permeable formation (modified and redrawn after Rider, 1996).

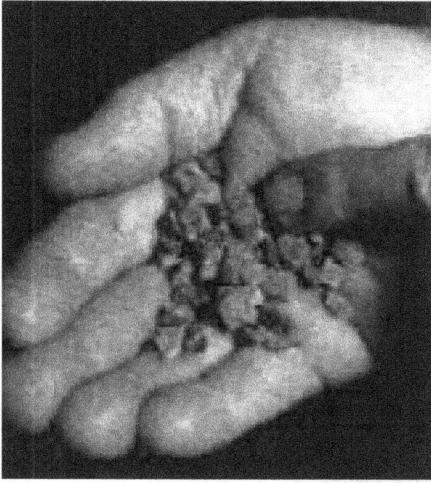

Figure 8.4. Cuttings produced by drilling (http://www.geomore.com/well-samples/).

the development of three concentric zones with respect to the axis of the borehole, as illustrated in Fig. 8.5.

The invaded formation consists of a *flushed zone* (6−10 cm in depth), close to the borehole, surrounded by a *transition zone*, which, in turn, is surrounded by the *uninvaded* or *undisturbed* part of the bed where the original formation fluids remain uncontaminated by the mud filtrate. In an oil-bearing formation, between 70% and 95% of the oil is displaced from the flushed zone and the diameter of invasion $(d_j)$ varies from inches to feet, depending on the pressure differential, formation and mud properties. The distribution of the fluids in a permeable zone after invasion is shown in Fig. 8.6.

It should be emphasised that invasion affects only the porous and permeable zones; tight formations permit little or no invasion.

### 8.2.2.  *Open-hole log types*

Open-hole logs may be categorised on the basis of the rock properties they measure. Accordingly, they fall into the following categories:

- electric logs:
  - spontaneous potential (SP),
  - resistivity,

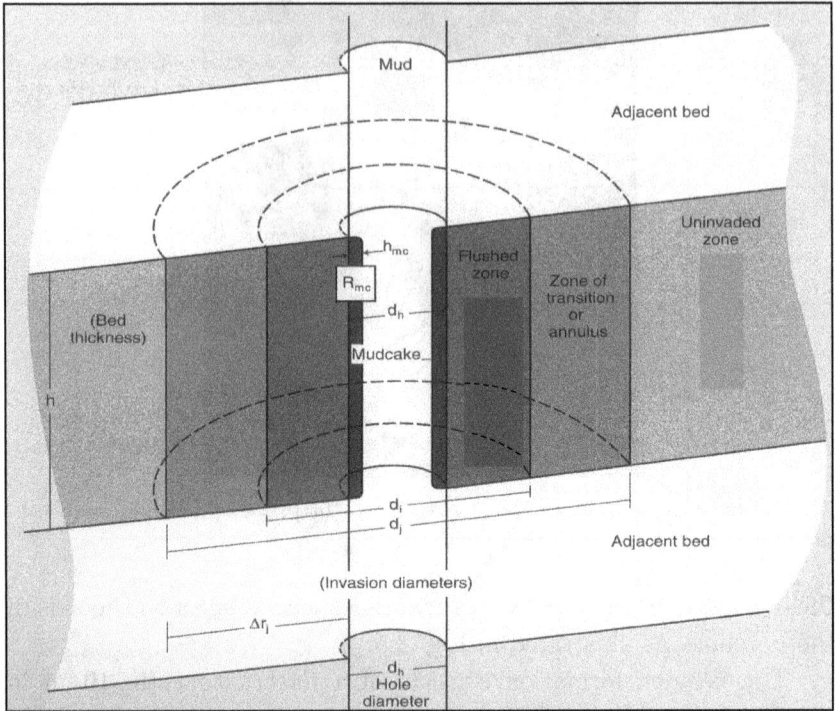

Figure 8.5. Invasion produces a cylindrical distribution of the displaced fluids around the well (modified image courtesy of Schlumberger, 2013).

- acoustic or sonic logs,
- radioactive or nuclear logs:

  o gamma ray,
  o neutron,
  o density,

- dielectric log,
- nuclear magnetic resonance (NMR) log,
- dipmeter and formation image logs.

### 8.2.3. *Terminology*

Extensive use of abbreviations and symbols is made in well logging. Broadly, these can be categorised into borehole parameters, drilling fluid parameters, formation characteristics, logging tools and log

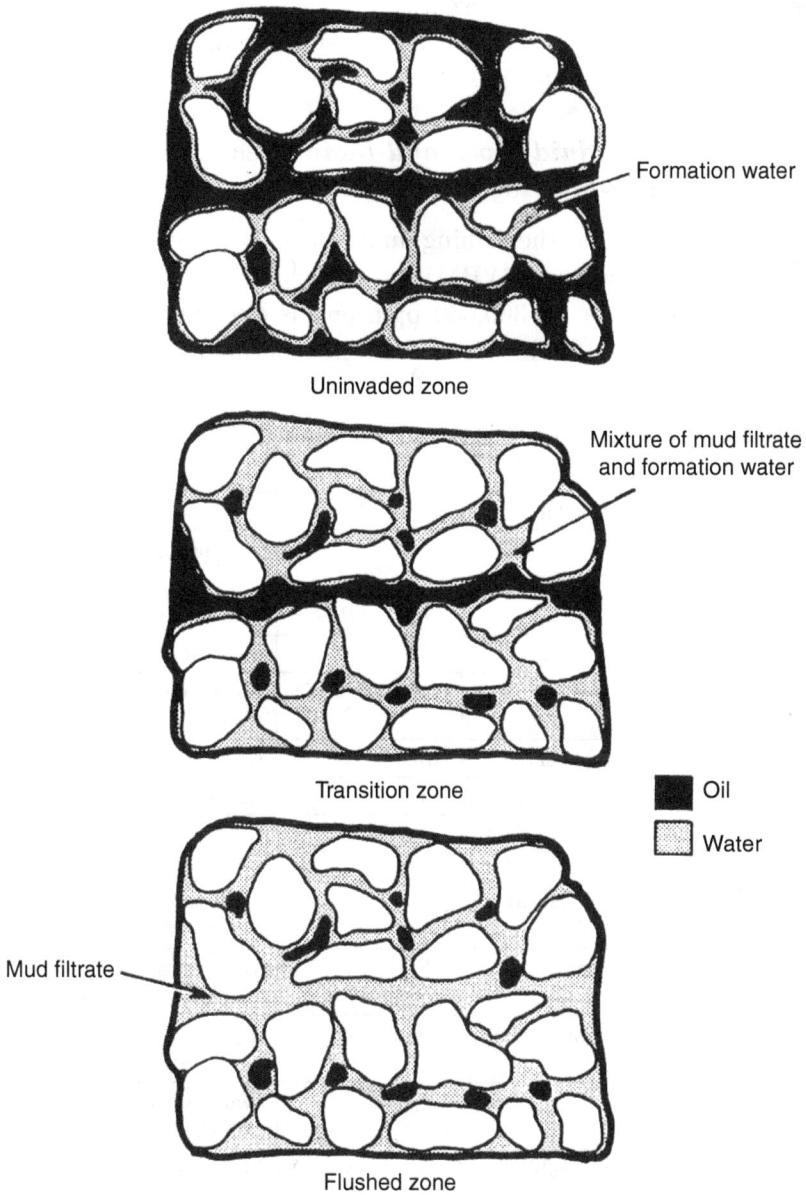

Figure 8.6. Distribution of fluids in a permeable zone after invasion (image courtesy of Sclumberger, 1987).

measurements. A detailed discussion of the nomenclature is beyond the context of this chapter but the most commonly used abbreviations are listed in Tables 8.1–8.4.

## 8.2.4. Drilling fluid types and their effect on open-hole log data acquisition

As mentioned earlier, the drilling fluid is usually a water- or oil-based mud. Water-based mud (WBM) has water as the continuous phase with salinities of up to 300,000 ppm and is electrically conductive. Oil-based mud (OBM) has oil as the continuous phase and is

Table 8.1. Formation characteristics and borehole parameters (Schlumberger terminology).

| $a$ | Tortuosity factor | $\Phi$ | Porosity |
|---|---|---|---|
| $BHT$ | Bottom hole temperature | $\Phi_A$ | Absolute porosity |
| $BS$ | Bit size | $Phi_E$ ($PHIE$) | Effective porosity |
| $BVW$ | Bulk volume water | $SPI$ | Secondary porosity index |
| $CALI$ | Caliper | $MOS$ | Moveable oil saturation $(S_{XO} - S_W)$ |
| $d_h$ | Diameter of borehole | $m$ | Cementation factor |
| $d_i$ | Diameter of flushed zone | $n$ | Saturation exponent |
| $d_j$ | Diameter of invaded zone | $ROS$ | Residual oil saturation $(1.0 - S_{XO})$ |
| $EFT$ | Estimated formation temperature | $S_h$ | Hydrocarbon saturation $(1.0 - S_w)$ |
| $F$ | Formation factor | $S_w$ | Water saturation of uninvaded zone |
| $h_{mc}$ | Thickness of mudcake | $S_{Wi}$ | irreducible water saturation |
| $k$ | Permeability | $S_{XO}$ | Water saturation of flushed zone |
| $k_A$ | Absolute permeability | $S_w/S_{XO}$ | Moveable hydrocarbon index |
| $k_Z$ | Effective permeability | $T$ | Formation temperature |
| $k_r$ | Relative permeability | $V_{sh}$ | Fractional volume of shale in formation |

Table 8.2. Electric logs (Schlumberger terminology).

| *AIT* | Array induction tool | *PL* | Proximity log |
|---|---|---|---|
| *CHFR* | Cased hole formation Resistivity tool | *R* | Resistivity |
| *DIL* | Dual induction laterolog | $R_{ILM}$ | Resistivity induction log medium |
| *DLL* | Dual laterolog | $R_{LLd}$ | Resistivity laterolog deep |
| *HRLA* | High resolution laterolog array | $R_{LLS_8}$ | Resistivity laterolog shallow |
| *IDPH* | Induction deep phasor | $R_{LLS}$ | Resistivity laterolog 8 |
| *IL* | Induction log | $R_m$ | Resistivity of drilling mud |
| *ILD* | Deep induction log | $R_{mc}$ | Resistivity of mudcake |
| *ILM* | Medium induction log | $R_{mf}$ | Resistivity of mud filtrate |
| *IMPH* | Induction medium phasor | $R_{MLL}$ | Resistivity microlaterolog |
| *LL* | Laterolog | $R_{MSFL}$ | Resistivity microspherically focused log |
| *LLD* | Deep laterolog | $R_O$ | Resistivity 100% water saturated formation ('wet resistivity') |
| *LLS* | Shallow laterolog | $R_S$ | Resistivity of adjacent beds |
| *LL8* | Laterolog 8 | $R_t$ | Resistivity of uninvaded zone |
| *LN* | Long (64″) normal | $R_W$ | Resistivity of formation water |
| *MINV* | Micro inverse curve | $R_{XO}$ | Resistivity of flushed zone |
| *ML* | Microlog | SN | Short (16″) normal |
| *MLL* | Microlaterolog | SP | Spontaneous potential |
| *MNOR* | Micro normal curve | *PSP* | Pseudostatic spontaneous potential |
| *MSFL* | Microspherically focused log | *SSP* | Static spontaneous potential |

electrically non-conductive. Most electric logs can only be run in the presence of a WBM since the signal requires a conductive medium to travel between the tool and the formations. This restricts the type and variety of the electric logs that can be obtained if an OBM

Table 8.3. Acoustic (sonic) log (Schlumberger terminology).

| $AST$ | Array sonic tool | $t(\Delta t; Dt)$ | Interval transit time of formation |
|---|---|---|---|
| $B_{cp}$ | Acoustic porosity compaction factor | $t_f$ | Interval transit time of fluid in formation |
| $BHC$ | Borehole compensated sonic log | $t_{ma}$ | Interval transit time of formation matrix |
| $LSS$ | Long spacing sonic log | | |

Table 8.4. Radioactive logs (Schlumberger terminology).

| $CGR$ | Computed gamma ray | $NPHI$ | Neutron porosity |
|---|---|---|---|
| $CNL/CNT$ | Compensated neutron log/tool | $P_e$ $(PEF)$ | Photoelectric absorption factor |
| $FDC$ | Formation density compensated Log | $\rho_b$ $(RHOB)$ | Bulk density of the formation |
| $GR$ | Gamma ray log | $\rho_f$ | Density of fluid in formation |
| $GR_{clean}$ | Gamma ray reading from clean zone | $\rho_{ma}$ | Density of the formation matrix |
| $GR_{shale}$ | Gamma ray reading from shale | $SGR$ | Standard gamma ray |
| $GR_{zone}$ | Gamma ray reading from formation | $SNP$ | Sidewall neutron porosity |
| $GST$ | Gamma ray spectrometry tool | $TDT$ | Thermal decay time log |
| $LDT$ | Litho-density tool | $TNPH$ | Total neutron porosity |
| $NGS$ | Natural gamma ray spectrometry log | $U$ | Volumetric photoelectric absorption index |

has been used. The nature of the drilling mud does not affect the acquisition of the other types of open-hole log data.

## 8.3.  Open-hole Log Interpretation

Open-hole log interpretation has two aspects: (a) qualitative and (b) quantitative and these are described briefly below.

### 8.3.1. *Qualitative interpretation*

The objectives of qualitative interpretation may be summarised as follows:

- identification of porous and permeable beds and their boundaries;
- identification of the pore fluids;
- determining the positions of fluid contacts (GOC, OWC and GWC); and
- correlation of subsurface strata.

### 8.3.2. *Quantitative interpretation*

The aims of quantitative interpretation include:

- quantification of porosity ($\Phi$) and permeability ($k$);
- calculation of water saturation, $S_W$, in the uninvaded part of a hydrocarbon-bearing zone, from which oil or gas saturation ($S_h$) may be deduced: $S_h = 1 - S_W$.
- calculation of water saturation, $S_{XO}$, in the flushed part of a hydrocarbon-bearing zone, from which residual oil or gas saturation, $S_{or}$, may be deduced: $S_{or} = 1 - S_{XO}$. A comparison of $S_W$ and $S_{XO}$ will provide an indication of the moveable oil saturation (MOS) in a hydrocarbon-bearing zone;
- estimation of the fractional volume of shale ($V_{Sh}$) in a permeable zone. This is necessary since all log readings are affected by the presence of shale, for which corrections must be made before using the values in calculations.

## 8.4. Open-hole Log Measurements and their Applications

Open-hole log measurements are an indispensable source of information on the physical properties of sub-surface rock layers and their interstitial fluids, a discipline referred to as *petrophysics*. These applications are discussed briefly below.

### 8.4.1. *Electric logs*

The SP log measures the difference in potential resulting from the movement of ions between conductive fluids, or *electrolytes*, in invaded formations. Formation waters and the mud filtrate, where WBM has been used as the drilling fluid, are the electrolytes and the ions, primarily sodium ($Na^+$) and chlorine ($Cl^-$) atoms, move from the more concentrated to the less concentrated electrolyte. Deflections of the SP curve identify permeable beds and provide information on formation water characteristics.

Resistivity logs record the resistance offered by the formations to the passage of an electric current. Resistivity, abbreviated to $R$, is a fundamental property of materials and is the reciprocal of conductivity, $C$:

$$R = 1/C. \tag{8.1}$$

The formation water is the only conductive element; hydrocarbons and dry reservoir rocks are insulators. Deep resistivity measurements (i.e. those from the uninvaded part of the formation) are the most important qualitative and quantitative indicators of hydrocarbons. Modern resistivity logs simultaneously record three to five curves of varying depth of investigation, broadly corresponding to the flushed, transition and uninvaded zones in a permeable bed. Since these zones contain different fluids (Fig. 8.5), their resistivity values will be different. Consequently, the curves separate and the separation is indicative of a permeable layer. Figure 8.7 presents an electric log suite, showing permeable and impermeable formations. The resistivity curve separation and the SP deflection associated with the interval 5,870–5,970 ft indicate that the layer is permeable. Furthermore, the low resistivity value recorded by the uninvaded zone curve, RILD, means that the formation contains a conductive fluid, i.e. saline water, and not hydrocarbons.

Impermeable zones remain uninvaded and are characterised by the resistivity curves tracking each other.

Hydrocarbon-bearing formations are normally identified by high uninvaded zone resistivity values as exemplified in Fig. 8.8.

Resistivity measurements also indicate hydrocarbon–saline water contacts. This is illustrated diagrammatically as well as in an actual case in Fig. 8.9.

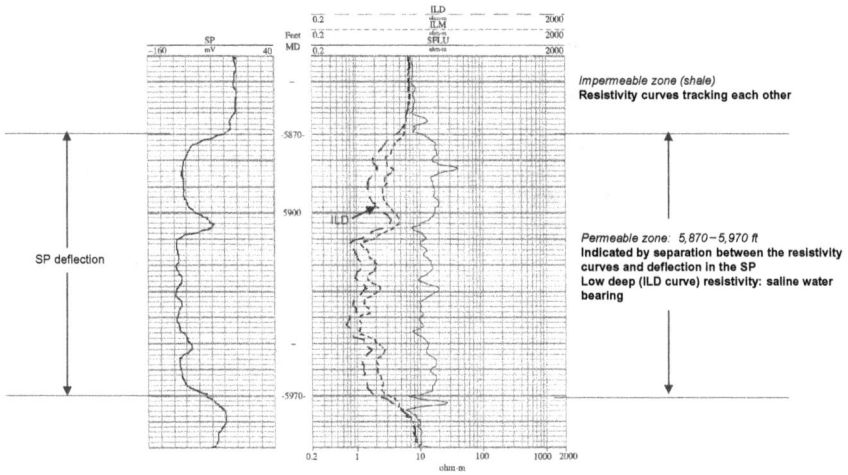

Figure 8.7. A suite of electric logs showing permeable and impermeable formations (from Asquith and Krygowski, 2004).

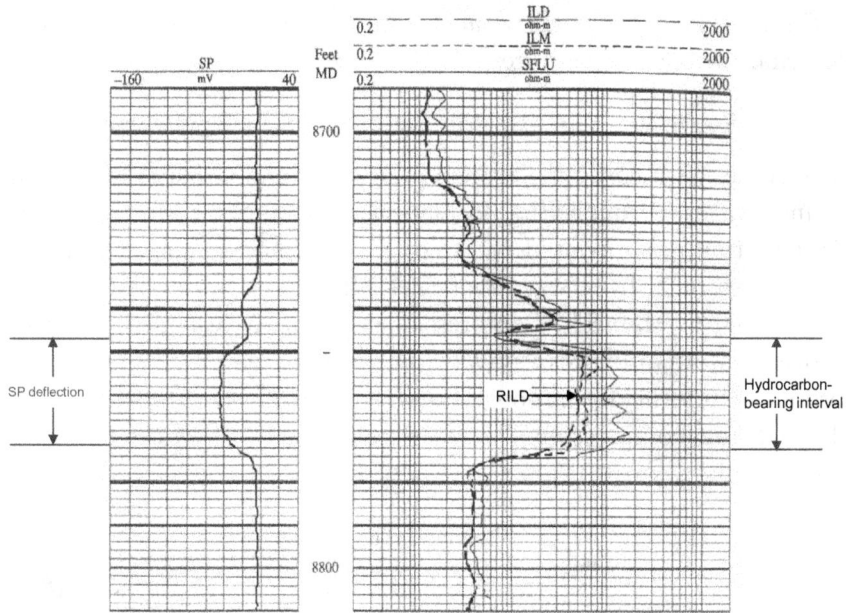

Figure 8.8. A suite of electric logs showing a hydrocarbon-bearing interval (from Asquith and Krygowski, 2004).

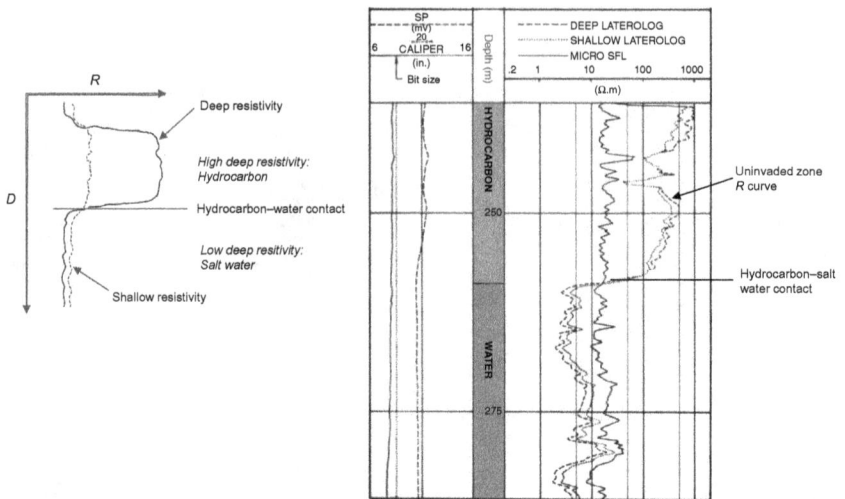

Figure 8.9. Identification of a hydrocarbon–saline water contact (images courtesy of Schlumberger, 1979).

It should be noted that deep resistivity becomes ineffective in differentiating hydrocarbons and fresh formation water, since both are characterised by high resistivity values.

### 8.4.2.  *Sonic or acoustic log*

The sonic log records the transit times of compressional and shear sound waves through the formations surrounding the borehole. Transit time, abbreviated to $\Delta t$ or $t$, is defined as the time taken in microseconds by a wave to travel 1 ft or 1 m through the layers penetrated by the well. Compressional and shear wave travel times provide information respectively on porosity and rock mechanical properties of the formations. Porosity of a given interval can be derived from compressional wave arrival times by using the following relationships:

$$\Phi = (t - t_{ma})/(t_f - t_{ma}), \tag{8.2}$$

$$\Phi = 0.63(1 - t_{ma}/t), \tag{8.3}$$

where $t$ is the interval transit time of the zone of interest ($\mu s/ft$ in the English system and $\mu s/m$ in the metric system); $t_{ma}$ is the interval

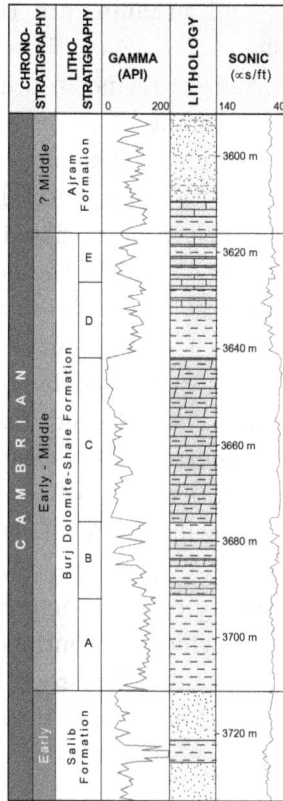

Figure 8.10. Example of a sonic log (after Aqrawi *et al.*, 2010).

transit time of the rock matrix (sandstone, limestone or dolomite — the common reservoirs); and $t_f$ is the interval transit time of the fluid in the pore spaces.

An example of a sonic log is presented in Fig. 8.10.

### 8.4.3. *Radioactive logs*

Radioactive logs provide information on the lithology and porosity of the rocks and they can be run in the presence of non-conductive muds. The gamma ray (GR) log measures the natural gamma radiation emitted by sedimentary rocks, while the neutron and density tools record the effects of artificial bombardment by nuclear particles.

Potassium, thorium and uranium are the sources of gamma radiation and occur in shales. Consequently, shales are characterised by high GR log values (Fig. 8.10) and in other rock types the GR readings increase as the content of shale increases. The GR log therefore reflects the proportion of shale in sediments.

The neutron and the density logs are useful indicators of lithology and the common reservoir rocks are associated with specific neutron-density log responses or *signatures*. In the case of sandstones, the neutron curve lies to the right of the density, while limestones are indicated by the two curves tracking each other with little or no separation. Dolomites are characterised by the density curve lying to the right of the neutron. These distinct signatures are the result of the way in which the neutron and density logs are calibrated. In practice, a GR-neutron-density log combination is the best approach to lithology determination. This is demonstrated in Fig. 8.11.

Reliable values of porosity can also be derived from the neutron and density logs. The latter, which is a continuous recording in g/cm$^3$ of the bulk density, $\rho_b$, of the subsurface rocks, is a particularly good indicator of porosity that can be derived through the application of the following relationship:

$$\Phi = (\rho_{ma} - \rho_b)/(\rho_{ma} - \rho_f), \qquad (8.4)$$

where $\rho_b$ is the density of the zone of interest (g/cm$^3$); $\rho_{ma}$ is the density of the rock matrix (sandstone, limestone or dolomite — the common reservoirs); and $\rho_f$ is the density of the fluid in the pore spaces.

It is noteworthy that the sonic, GR, neutron and density logs are shallow reading tools; their depth of investigation is confined to the flushed zone. This does not, however, affect the validity of the interpretation results, since rock properties, i.e. lithology and porosity, do not change from the flushed zone through the transition zone to the uninvaded part of the formation.

### 8.4.4.   *Dielectric and NMR logs*

Dielectric and NMR logs are relatively recently introduced logging devices and the principles of their operation and data interpretation

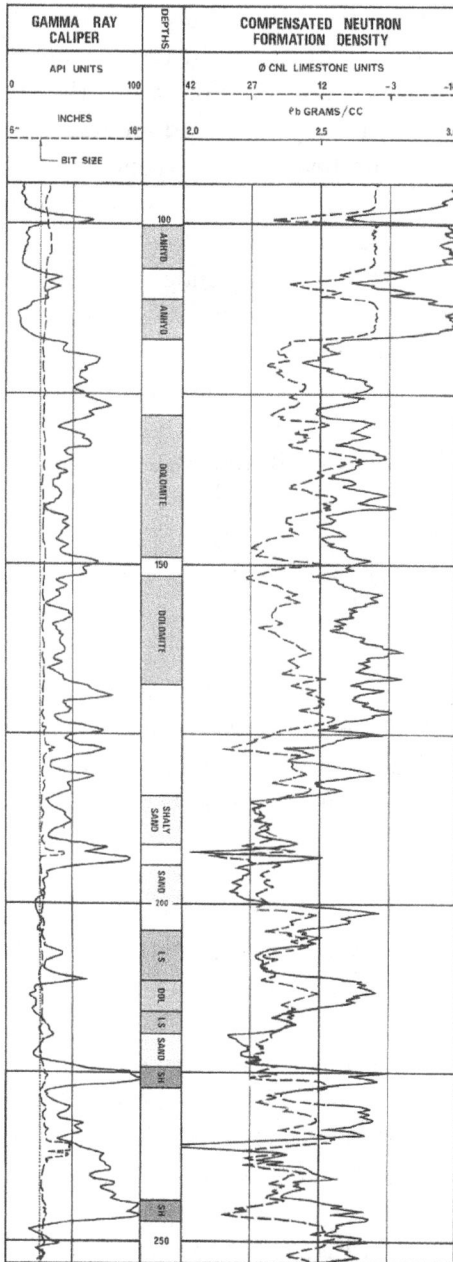

Figure 8.11. Lithology determination by the GR-neutron-density log combination (image courtesy of Schlumberger, 1976).

are complex. The dielectric log measures the transit time of electro-magnetic waves, which is influenced by the formation dielectric properties. Since the waves travel more or less at the speed of light, the measured times are extremely short and are recorded in nanoseconds per meter, a nanosecond being $1/10^9$ second. Its main application is the detection of water regardless of its salinity, offering a means of identifying hydrocarbons in freshwater-bearing formations.

The NMR is a lithology independent tool and responds to the presence of hydrogen in the formations. It is a very versatile log and provides a large variety of information on rock and fluid properties, including the identification of various types of water, porosity and permeability and presence of hydrocarbons. It is particularly useful in complex lithology and shaly reservoirs where, in the case of the latter, certain calculations based on standard log data can produce spurious results. An interesting feature of the device is that, although it makes a nuclear measurement, it does not employ a nuclear source.

### 8.4.5.  *Dipmeter*

The dipmeter provides a continuous measurement of the inclination and dip direction of the formations penetrated by the borehole. Determinations of the angle and direction of dip are usually presented as a plot of 'tadpoles' versus depth as shown in Fig. 8.12. The head of the tadpole indicates the dip angle and its tail points to the azimuth (dip direction).

Dipmeter data have applications in structural and sedimentary geology. In structural geology the data are used in the interpretation of unconformities, folds and faults, while in sedimentary geology the dipmeter can assist in facies analysis and determination of depositional environments. It must be emphasised, however, that dipmeter data should never be interpreted alone and need to be integrated with other standard logs since different dipmeter log patterns often indicate the same geological phenomena. Four patterns are recognised in dipmeter log motifs: green, red, blue and random. These are illustrated in Fig. 8.13.

*Green pattern* is characterised by constant dip and azimuth with depth and usually represents post depositional structural dip. It is typical in shale sequences.

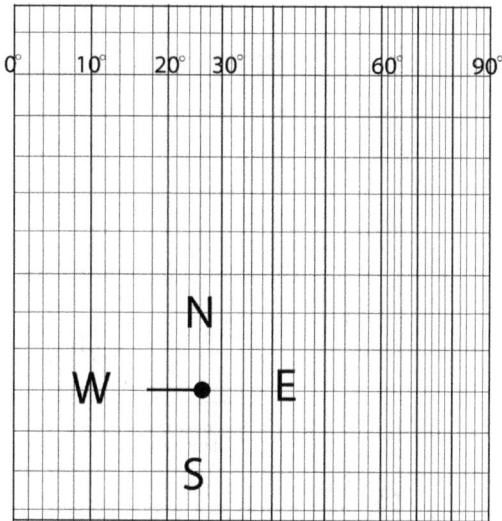

Figure 8.12. Dipmeter log presentation. Standard presentation is a recording of 'tadpoles' versus depth.

*Red pattern* represents uniform azimuth but the dip angle decreases upward. This pattern is usually associated with folds, faults, unconformities, reefs, channel sands and valley fills.

*Blue pattern* exhibits roughly uniform azimuth but an upward increase in dip angle. This pattern is usually associated with folds or faults, unconformities, palaeocurrent directions and submarine fan deposits.

*Random pattern* shows scattered dip and azimuth. Massive beds lacking coherent bedding planes or slumped deposits may be responsible for such a response. It may also result from tool malfunction or poor borehole conditions causing the pads to lose contact with the borehole wall.

### 8.4.6. *Formation image logs*

Formation image logs provide computer-generated images based on the measurement of the resistivity or the acoustic reflectivity of the formations. They resemble core photographs and their acquisition is much more cost-effective but they do not obviate the necessity of taking cores. Core samples are required for laboratory

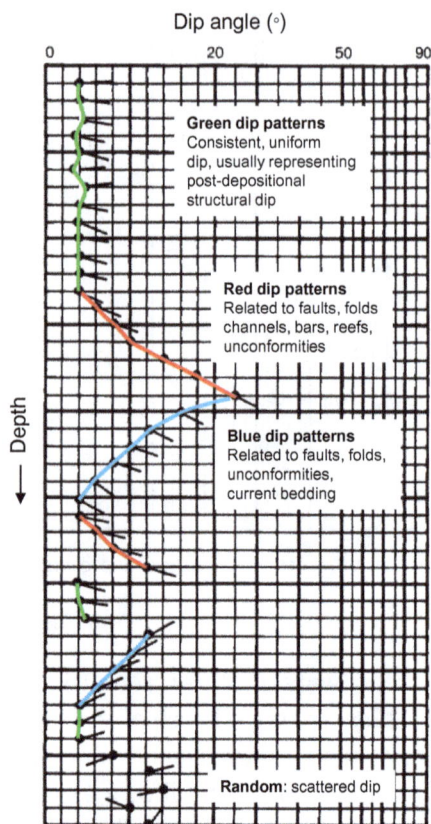

Figure 8.13. Coloured dip patterns and their possible interpretations.

determinations — referred to as *core analysis* — of the petrophysical and other properties of the subsurface rocks. Figure 8.14 presents an image log and a scanned image of its corresponding core section.

## 8.5.   Summary of the Applications of Open-hole Log Interpretation Results

Permeable zone indicators:

- Generally low GR, unless shaly,
- Presence of mudcake (resulting from invasion and detected from the caliper log),

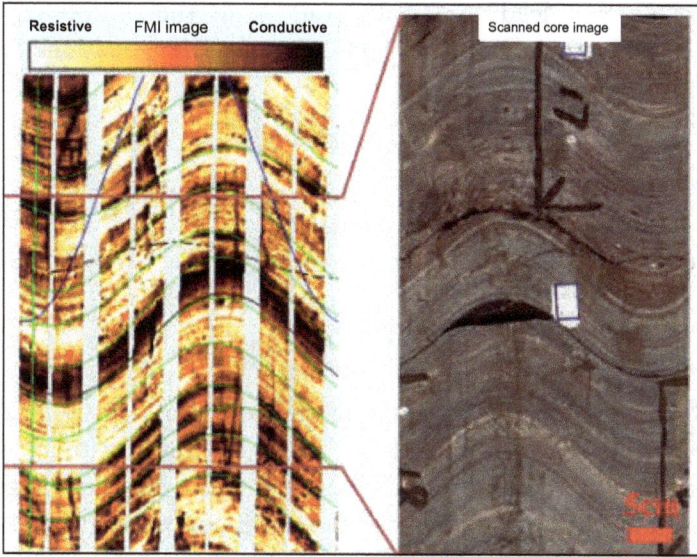

Figure 8.14. An image log and a scanned image of its corresponding core section (from http://www.intechopen.com).

- Resistivity curve separation,
- SP deflection,
- Indication of porosity from the sonic, neutron and density logs,
- Indications from NMR log measurements.

Impermeable zone indicators:

- Generally high GR (except evaporites),
- No mudcake (no invasion),
- No resistivity curve separation,
- No SP deflection,
- No indication of porosity from the sonic, neutron and density logs.

Hydrocarbon indicator:

- High deep resistivity.

### 8.5.1. *Basic quantitative open-hole log interpretation*

Determination of $S_W$ in a hydrocarbon-bearing reservoir is the most important objective of quantitative interpretation. As discussed earlier, $S_W$ will lead to a value for $S_h$ in the uninvaded zone, an essential parameter in the STOIIP and GIIP equations for the estimation of reserves:

$$\text{STOIIP} = [\text{GRV} \times N/G \times \Phi(1 - S_w)]/B_o, \qquad (8.5)$$

$$\text{GIIP} = \text{GRV} \times N/G \times \Phi(1 - S_w) \times B_g. \qquad (8.6)$$

Measured deep resistivity, $R_t$, is related to $S_W$ by the following relationship:

$$R_t = (aR_W)/(\Phi^m S_W^n), \qquad (8.7)$$

where $a$ is an intrinsic property of the reservoir depending on its lithology and degree of consolidation; $R_W$ is the resistivity of the formation water in the uninvaded zone; $\Phi$ is the porosity (fraction); and $m$ and $n$ are variables whose most commonly used value is 2.

Solving Eq. (8.7) for $S_w$ and assuming $m = n = 2$:

$$S_w^2 = (aR_W)/(\Phi^2 R_t). \qquad (8.8)$$

Therefore,

$$S_w = \sqrt{(aR_W)/(\Phi^2 R_t)}. \qquad (8.9)$$

Equation (8.9) remains valid as long as the shale content ($V_{sh}$) of the formation is below 15−20%. In the presence of higher shale volumes, the $S_w$ value derived from Eq. (8.9) will be an overestimate, since shale lowers $R_t$, causing a reduction in $S_h$, as $S_h = 1 - S_w$. This, in turn, will result in an underestimation of the reserves. In such cases different and more complex equations must be used to discount the effect of shale.

### 8.5.2.  *Computer processing and interpretation*

Computer processing and interpretation techniques are used extensively in the qualitative and quantitative interpretations of open-hole log data. A wide range of interpretation software is available that produces continuous curves displaying rock and fluid properties. An example of a computer processed log interpretation is presented in Fig. 8.15.

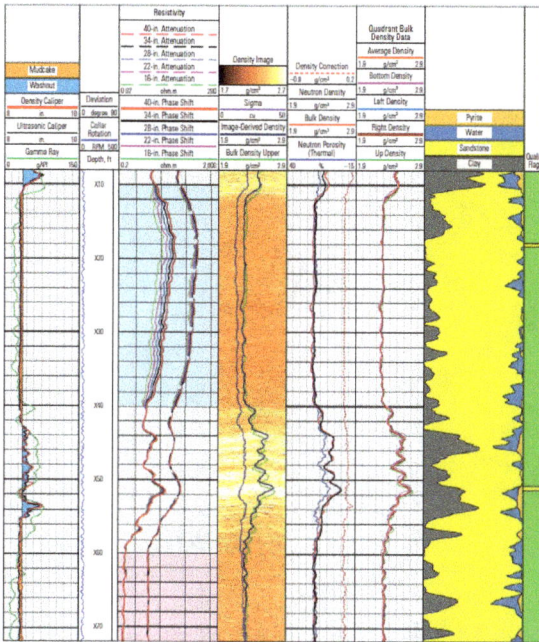

Figure 8.15. Examples of computer-processed log interpretation (image copyright Schlumberger, used with permission and courtesy of Schlumberger).

# References

Aqrawi, A.M.A., Goff, J.C., Horbury, A.D. *et al.* (2010). *The Petroleum Geology of Iraq*, Scientific Press, UK.

Asquith, G. and Krygowski, D. (2004). *Basic Well Log Analysis, 2$^{nd}$ Edition*, AAPG Methods in Exploration Series 16, AAPG, Tulsa, OK.

Rider, M.H. (1996). *The Geological Interpretation of Well Logs: 2$^{nd}$ Edition*, Gulf Publishing Corp., Houston.

Schlumberger (1976). Well Evaluation Conference, Iran.

Schlumberger (1979). Well Evaluation Conference, Algeria.

Schlumberger (1987). Basic Log Interpretation Seminar, UKI Division.

Schlumberger (2013). Log Interpretation Charts.

# Glossary of Technical Terms
## and Abbreviations

**A**

**AAPG.** American Association of Petroleum Geologists. Founded in 1917, it is the world's premier professional association for petroleum explorationists. It has more than 40,000 members worldwide and is dedicated to the advancement of the science of petroleum geology through publications, conferences, educational courses and by promoting research.

**Absolute permeability.** Value of permeability measured when only one fluid is present and it fully saturates the rock. In this case, the permeability of the rock to that fluid is a maximum and is called its absolute permeability, abbreviated to $k_A$.

**Absolute porosity.** The ratio between total pore volume (regardless of whether or not the pores are interconnected) and bulk volume in a rock. It is abbreviated to $\Phi_A$.

**Absolute time.** The numerical age of a geological feature or event specified in years.

**Accommodation space.** Space made available by subsidence for sediment accumulation in a sedimentary basin.

**Accreting plate margin.** A margin where new crust is created by the upwelling of magma from the mantle and added to the crust. These margins coincide with the mid-ocean ridges and result in the spreading of the ocean floor. Also referred to as *constructive* or *divergent* plate margins.

**Acoustic impedance.** Parameter that determines the reflective properties of the interface between two layers. It is abbreviated to $I_a$ and defined as the product of the density, $\rho$, and compressional wave velocity, $V_p$, of the layers separated by the interface: $I_a = V_p \times \rho$.

**Alkanes.** Saturated paraffins, characterised by a straight chain molecular structure and single carbon-to-carbon bonds. They have the general formula $C_nH_{2n+2}$.

**Alkenes.** Unsaturated straight chain hydrocarbons, containing at least one carbon-to-carbon double bond and having the general formula $C_nH_{2n}$. Also known as *olefins*.

**Alluvial fan.** A gently sloping apron of sediments built by streams at the base of a mountain in arid or semi-arid regions.

**Anhydrite.** See *evaporite*.

**Anoxic environment.** Lack of oxygen at the sediment–water interface, a condition necessary for the preservation of the organic matter in source rocks.

**Anticline.** An arch in which the two sides, usually called limbs or flanks, dip away from one another. The strata forming the centre of the fold are referred to as the 'core' and those forming the outer part are called the 'envelope'. In an anticline, the 'core' is older than the 'envelope'.

**API.** American Petroleum Institute. Established in 1919, it is a national trade association that represents all sectors of America's oil and natural gas industries. It has more than 600 corporate members which include producers, refiners, suppliers, pipeline operators and marine transporters as well as service and supply companies that support the oil and gas industry. It collects, maintains and publishes statistics and data on all aspects of US industry operations and leads the development of petroleum and petrochemical equipment and operating standards. Although its focus is primarily domestic, in recent years its work has expanded and now includes an international dimension.

**API gravity.** A measure of the density of crude oil according to a scale devised by the American Petroleum Institute.

The measurement is expressed in API and is inversely proportional to specific gravity.

**Appraisal well.** A well drilled to confirm and evaluate the presence of hydrocarbons in a reservoir that has been discovered by an exploration or wildcat well.

**Aromatics.** Unsaturated closed ring hydrocarbons with the general formula $C_nH_{2n-6}$.

**Asthenosphere.** The relatively plastic layer, underlying the *lithosphere*, on which the tectonic plates move. It is also known as the *upper mantle* and is approximately 200 km thick.

**Atomic number.** The number of protons (positively charged particles) in the nucleus of a given element.

**Atomic weight.** The number of protons plus the number of neutrons (particles with no electric charge) in the nucleus of a given element.

**Azimuth.** Direction of the dip of a surface measured as a clockwise rotation from the north.

# B

**BCF.** Billion cubic feet.

**BCM.** Billion cubic metres.

**BHT.** Bottom hole temperature, the maximum temperature measured and recorded in a well.

**Biochemical rocks.** Commonly carbonates formed by the life processes of living organisms such as corals and algae.

**Biofuel.** Petrol and diesel extracted from plants.

**Breccia.** A coarse grained clastic sedimentary rock made up of angular fragments held together by a fine grained matrix.

# C

**Calcite.** Calcium carbonate ($CaCO_3$), the dominant constituent of *limestones*.

**Cap rock.** See *Seal*.

**Casing.** Strings of steel pipe cemented into oil and gas wells. Each string is cemented all the way to the surface. It protects the borehole wall, prevents it from collapse in unconsolidated formations and shuts off flow of unwanted fluids into the well. Hydrocarbons are produced through perforations in the production section of the string.

**Chalk.** A biochemical carbonate made up entirely of the remains of primitive, submicroscopic algae known as *coccoliths.*

**Clastic rocks.** Sedimentary rocks formed by the consolidation of individual particles or grains derived from the erosion/breakdown of pre-existing rocks and minerals. They are not *in situ* deposits since their formation involves transportation of the grains or fragments from their source to a place of deposition through the agencies of water, wind and ice. *Sandstones* and *shales* are examples of clastic rocks.

**Closure.** Vertical distance between the crest of the structure and spill point. Closure is an important feature of an anticlinal trap — it controls the height of the hydrocarbon column and the storage capacity of the structure.

**Coal bed methane.** Abbreviated to CBM, it includes methane generated during coal formation and adsorbed in coal. It is produced by drilling into coal seams, decreasing the pressure which allows the gas to bubble out.

**Combination trap.** Trap formed by a combination of both structural and stratigraphic factors.

**Compaction.** Reduction in the volume of a deposit due to the expulsion of water caused by increasing overburden pressure during burial.

**Condensate.** Hydrocarbon occurring as gas under reservoir temperature and pressure but as liquid under surface conditions.

**Conglomerate.** A coarse grained clastic sedimentary rock composed of rounded fragments larger than 2 mm in diameter bound together by a matrix of sand and silt.

**Continental crust.** The type of crust that underlies the continents and continental shelves.

**Continental shelf.** The area at the edges of a continent from the shoreline to a depth of 200 m, where the *continental slope* begins. The shelf is a generally flat region with a slight seaward slope.

**Continental slope.** The relatively steep slope that separates the *continental shelf* from the deep ocean basin.

**Contour.** A line that joins points of equal value at the surface or in the subsurface.

**Conventional resources.** Hydrocarbons occurring in pores in the *reservoir rock* from which they are extracted by drilling production wells.

**Core.** Cylindrical sample of rock, usually of the reservoir, cut during drilling.

**Correlation.** Demonstration of equivalency, involving the identification of certain characteristics in one well and recognition of the same features in a second or subsequent wells. The intervals exhibiting the same features are considered to be equivalent and can be matched.

**Crest.** The highest point of an anticline.

**Critical moment.** The time of hydrocarbon migration. This must predate, or at least be simultaneous with, trap formation.

**Crust.** See *lithosphere*.

**Cycloalkanes.** Saturated closed ring hydrocarbons with the general formula $C_nH_{2n}$. Also referred to as *cycloparaffins* or *naphthenes*.

## D

**2D seismic survey.** Data acquisition using 2–3 km grid spacing. Provides regional coverage.

**3D seismic survey.** Data acquisition using 20–30 m grid spacing.

**4D seismic survey.** At least two 3D surveys run at different times over the same area, usually a field. The '4th dimension' is time.

**Daughter element.** Isotope produced by radioactive decay.

**Deep sea trench.** Long, narrow, steep-sided depression in the ocean floor, coinciding with destructive plate margins and earthquake belts.

**Delineation well.** A well drilled in an existing field to determine the extent of the reservoir.

**Depocentre.** The area of greatest sediment thickness at any given time in a sedimentary basin.

**Destructive plate margin.** A margin at which the crust formed at the accreting margins descends into the mantle and is destroyed. These margins coincide with deep sea trenches and earthquake belts. Also known as *consuming* or *convergent* plate margins or *subduction zones*.

**Development well.** A post-discovery well drilled to the depth of the hydrocarbon-bearing layer. Development wells become *production wells* once the field goes on stream.

**Diagenesis.** The processes that are involved in the conversion of soft sediments into lithified *sedimentary rocks*.

**DHI.** Direct hydrocarbon indicator.

**Dip.** The maximum inclination of a surface, normally a bedding or a fault plane, from the horizontal.

**Dolomite.** Calcium-magnesium carbonate [Ca Mg $(CO_3)_2$], the dominant constituent of dolomite or dolostone.

**Dolomitisation.** The process by which *calcite* is converted into dolomite.

**Dolostone.** See *dolomite.*

**Dyke.** An intrusive igneous rock that invades pre-existing rocks, forming vertical or sub-vertical bodies. Unconsolidated sedimentary rocks can also intrude into pre-existing rocks and form dykes.

**E**

**Earthquake.** Sudden release of accumulated stress in the Earth's crust by movement or violent shaking.

**Economic basement.** The rocks below the sedimentary section in a sedimentary basin. Usually igneous and metamorphic rocks; non-petroliferous.

**Effective permeability.** When a rock contains more than one fluid, as is the case in most reservoirs, its permeability to any one fluid is reduced. The ability of a reservoir to conduct one fluid in the presence of others depends on the saturation of that fluid and is called its effective permeability, abbreviated to $k_E$. $k_E$ changes as the saturation of the fluid in question varies.

**Effective porosity.** Abbreviated to $\Phi_E$, and defined as the ratio between the interconnected pore spaces and bulk volume. $\Phi_E$ is an important factor since the permeability of the rock depends on its effective porosity.

**EI.** Energy Institute, created in 2003 by the merger of two organisations: the Institute of Petroleum (IP) and the Institute of Energy (InstE), dating from 1914 and 1927, respectively. The former was founded in 1914 as the Institution of Petroleum Technologists and renamed the Institute of Petroleum in 1938, and the latter was established as the Institute of Fuel in 1927, changing its name to the Institute of Energy in 1978. The EI has 14,000 members worldwide and publishes a Yearbook and Directory of organisations and consultancies operating in the energy sector as well as a quarterly (*Journal of the Energy Institute*) and two monthly (*Petroleum Review* and *Energy World*) magazines, covering a broad spectrum of issues including business news, technology development, international energy policies, energy sustainability, safety codes, industry standards, carbon management and climate change. It also provides library and educational services.

**EIA.** US Energy Information Administration. Established in 1977, EIA is part of the US Department of Energy responsible for collecting and disseminating energy statistics and analysis. The EIA conducts a comprehensive data collection programme that covers the full spectrum of energy sources, end uses and generates short- and long-term domestic and international energy projections. It performs informative energy analyses and disseminates its data products, reports, and services to customers and stakeholders primarily through its website and the customer contact centre. In 2015, it had 370 employees and a budget of $117 m.

**EOR.** Enhanced oil recovery.

**Evaporites.** Sedimentary rocks precipitated from concentrated sea water. The most common types include rock salt or halite ($NaCl$), potassium salts (primarily sylvite, $KCl$), anhydrite ($CaSO_4$) and gypsum ($CaSO_4$ $2H_2O$).

**Exploitation well.** Synonymous with *production well.*

**Exploration well.** A well drilled in search of an as-yet-undiscovered oil or gas accumulation.

**F**

**Facies.** A distinctive group of features (rock type or lithology, colour, mineral composition, fossil content) that characterise a rock unit. Facies represent the record of a sedimentary environment which existed in the geologic past but is no longer present.

**Facies map.** A map that shows lateral variations in facies or rock type of a given layer.

**Fault.** A fracture or dislocation along which there has been movement on one side relative to the other.

**Fault plane.** The surface along which movement has taken place.

**Fault throw.** The amount of vertical movement that has occurred along a fault plane.

**Fauna.** The animal fossil remains found in some sedimentary rocks.

**Field.** A commercially viable conventional accumulation of oil, gas or condensate.

**Flora.** The plant fossil remains found in some sedimentary rocks.

**Fluvial.** Processes associated with rivers and streams and the deposits and landforms created by them.

**Footwall.** The rocks below an inclined fault plane.

**Formation.** A rock unit which is different from the layers above and below by its lithological characteristics — mineralogy, colour, grain size, texture and structures — and has regional lateral continuity.

**Formation volume factor.** A parameter, abbreviated to $B_O$ for oil and $B_g$ for gas, used to convert hydrocarbon volumes from reservoir to stock tank (standard or surface) conditions. It is the ratio between the volume of hydrocarbons at reservoir pressure and temperature to that at standard or surface conditions. See also under *PVT analysis*.

**Fracking.** The process of pumping a mixture of water, chemicals and sand under high pressure into tight formations and fracturing them to induce permeability and allow the release and production of hydrocarbons.

## G

**Gas cap expansion drive.** Production mechanism where the expansion of the gas cap, as the result of a pressure drop in the reservoir, causes the GOC to move downward, driving the oil into the production wells and ultimately to the surface.

**Gas hydrate.** A naturally occurring substance composed of molecules of methane trapped in a lattice of water molecules. It is stable at high pressure and low temperature and physically resembles ice.

**Gas window.** The depth range on the time–temperature curve where gas is generated from kerogen.

**Gather.** Ray paths from a single source point shooting into multiple receivers.

**Geology.** The study of rocks and earth history.

**Geophone.** A receiver used in a seismic land survey.

**GIIP.** Gas initially in place.

**GOC.** Gas–oil contact.

**GOR.** Gas–oil ratio. The volume of gas that comes out of solution as pressure drops when oil reaches the surface from the reservoir. It is measured in standard cubic feet of gas per barrel of oil.

**Graben.** A downfaulted block.

**GRV.** Gross rock volume, product of hydrocarbon-bearing area (acres or hectares) and gross mean reservoir thickness (feet or metres).

**GSL.** Geological Society of London. Founded in 1807, it is the oldest learned organisation of its kind in the world. It is the UK's professional body for Earth science and has a worldwide membership of over 11,500. The Society promotes the Earth sciences through the publication of books and journals, library and information services, cutting-edge scientific conferences, education activities and accreditation of under-graduate and postgraduate degree programmes.

**GWC.** Gas–water contact.

**Gypsum.** See *evaporite.*


# H

**Half graben.** A rotated graben.

**Half-life.** The time taken for 50% of the parent atoms to decay into daughter atoms.

**Hanging wall.** The rocks above an inclined fault plane.

**Horst.** An upfaulted block.

**Hydrocarbons.** Compounds composed of hydrogen and carbon atoms. The term is often used interchangeably with petroleum.

**Hydrocarbons in place.** Total volume of oil or gas that is present in a field before any production.

**Hydrophone.** A receiver used in a marine seismic survey.


# I

**IEA.** The International Energy Agency. An organisation founded in 1974 by oil and gas consuming nations. Currently, it has 29 member countries and its function is to examine the full spectrum of energy supply and demand and advocate policies that will enhance the reliability, affordability and sustainability of energy in its member countries as well as globally. The organisation produces many publications focusing on statistics and analysis of energy-related issues.

**Igneous rocks.** Formed by the solidification of magma or molten material that rises from the interior of the Earth. These are the 'parent' rocks of the other two groups.

**Injection well.** A well through which fluids are injected into a reservoir to increase pressure and to displace hydrocarbons.

**Inner core.** The innermost part of the Earth. It is a solid ball with a radius of 1,220 km, consisting of an iron–nickel alloy.

**Integrated oil company.** An enterprise that is involved in a significant level in *upstream* (exploration, field development and production), *midstream* (transportation by tankers and pipeline) and *downstream* (refining, marketing, distribution and sales) activities.

**IOC.** International oil company.

**IOR.** Improved oil recovery. Synonymous with *EOR*.

**Isolith map.** A map showing variations in the net thickness of a single rock type.

**Isopach map.** A map showing lateral variations in the thickness of a given layer.

**Isopermeability map.** A map showing lateral variations in the mean permeability of a reservoir rock.

**Isoporosity map.** A map showing lateral variations in the mean porosity of a reservoir rock.

**Isotope.** Different versions of an element that have the same atomic number but different atomic weights.

# K

**Kerogen.** The organic matter in sedimentary rocks that is insoluble in organic solvents. Hydrocarbons are generated by the thermal alteration of kerogen.

**Kitchen.** An area, normally located in the central parts of basins where the sediments have been buried to depths and subjected to temperatures sufficient to bring about source rock maturation and hydrocarbon generation.

# L

**Law of Superposition.** In a normal sedimentary sequence, i.e. one that is not upside down or inverted, the lowermost bed is always the oldest and topmost layer is always the youngest.

**Lead.** A possible target for exploration drilling, identified on the basis of regional geological study and/or 2D seismic data interpretation. It requires further evaluation before drilling.

**Limestone.** A carbonate rock made up of the mineral *calcite* ($CaCO_3$).

**Lithology.** Rock type, determined by its nature and mineralogical composition.

**Lithosphere.** The brittle outer layer of the Earth commonly referred to as the *crust*. It contains all the Earth's mineral deposits and fossil fuel resources but forms only 0.6% of the Earth's radius. It is divided into *oceanic* and *continental* types, with the former being denser. Lithospheric thicknesses vary from 40 km in continental areas to 10 km in oceanic regions.

**LNG.** Liquified natural gas — methane, $CH_4$.

**Log.** A continuous recording versus depth of a set of curves representing various properties of the formations penetrated in wells.

**LPG.** Liquified petroleum gas — propane, $C_3H_8$, or butane, $C_4H_{10}$.

**LWD.** Logging while drilling.

**M**

**Magma.** Molten material which forms in chambers below the Earth's surface. It can either rise to the surface as lava and form extrusive igneous rocks or cool within the Earth's lithosphere to form intrusive igneous rocks.

**Mantle.** The intermediate layer of the Earth beneath the crust. It is about 2,900 km thick and overlies the core of the Earth. The mantle consists of dense *igneous* rocks.

**Maturity (source rocks).** A stage in the thermal evolution of source rocks marking the onset of hydrocarbon generation from kerogen.

**Member.** A subdivision of a formation. Regional lateral continuity is not required and it represents local lithological variations in the formation.

**Metamorphic rocks.** Formed by the alteration of pre-existing rocks through an increase in temperature and pressure. The alteration occurs in the solid state, which means that it does not involve melting.

**Mid-ocean ridge.** Submarine mountain chains that are present in all of the world's oceans.

**Migration.** (a) Movement of oil and gas in the subsurface (see *primary* and *secondary* migration). (b) Relocating the reflections from a surface to positions where they should be in a seismic section.

**M.** 1,000 units.

**MM.** 1,000,000 units.

**Monocline.** A simple step-like flexure in which more or less horizontal beds locally assume a dip in one direction and then flatten out again.

**Mudstone.** A very fine grained clastic sedimentary rock composed of silt and clay size particles (less than 1/16 mm in diameter). Compositionally, it is the same as shale but differs from shale in being blocky or massive rather than laminated.

**MWD.** Measurement while drilling.

**N**

**NMO.** Normal move out, a term used in seismic data processing.

**Normal fault.** A fault along which the hanging wall has moved down with respect to the footwall. Caused by tension.

**N/G.** Net-to-gross ratio.

**O**

**Oceanic crust.** The type of crust that underlies the oceans.

**Offset.** The distance between a source and receiver in a seismic survey.

**OIIP.** Oil initially in place.

**Oil shale.** A fine grained sedimentary rock that produces oil on heating. Oil shales are immature source rocks.

**Oil sand.** Sand and clay heavily impregnated with highly viscous bitumen. Also known as tar sand.

**Oil window.** The depth range on the time–temperature curve where oil is generated from kerogen.

**Olefins.** See alkenes.

**Outer core.** A layer, about 2,300 km thick, between the Earth's inner core and mantle. It is believed to be liquid and composed of iron and some nickel.

**OWC.** Oil–water contact.

**P**

**Parent element.** A radioactive element that undergoes decay.

**Pay.** Hydrocarbon-bearing intervals encountered in a well.

**Permeability.** A complex quantity, representing the ability of the rock to transmit fluids and depends on the degree of connection between the pore spaces, i.e. on effective porosity. It is abbreviated to k.

**PESGB.** Petroleum Exploration Society of Great Britain. Established in 1964, it is a professional association with more than 4,800 corporate and individual members. It promotes education in the scientific and technical aspects of petroleum exploration through publications and by organising lectures, seminars and conferences.

**Petroleum events chart.** A 2D chart showing the distribution of the petroleum system elements in time and dating the generation and migration processes.

**Plate tectonics.** A concept, introduced into Earth science literature in 1967–1968, that the lithosphere is divided into a number of internally rigid blocks or plates of varying size that are in continuous motion relative to one another.

**Play.** A group of fields or drilling prospects in a given region identified by common geological and engineering characters. The former include common source, reservoir, seal, trap, timing, migration and preservation features, while the latter embrace the fluid properties of the hydrocarbons and flow characteristics of the producing reservoirs.

**Play fairway analysis (PFA).** Assessment of exploration risk at a basin or regional scale.

**Porosity.** The ratio between pore volume and bulk volume in a rock expressed as a per centage. It is abbreviated to $\Phi$ and is a measure of the storage capacity of the rock.

**Possible reserves.** Quantities of oil or gas expected from future discoveries in areas or formations known to be productive.

**Primary migration.** Expulsion of the newly generated hydrocarbons from their source rock.

**Principle of original horizontality.** The surfaces on which sediments accumulate — such as the floodplain of a river or the bed of a lake or sea — are approximately horizontal. A steep slope would cause the sediments to slide downslope before lithification (consolidation), which means they would not be preserved as sedimentary layers. Folds and tilted beds therefore indicate that the layers have been subjected to post-depositional deformation.

**Probable reserves.** Quantities of oil or gas in extensions of existing fields beyond or below the currently known limits.

**Production mechanism.** The natural process driving the hydrocarbons through and out of the reservoir into the production wells. Also known as *drive mechanism.*

**Production well.** A well through which hydrocarbons are produced.

**Prospect.** A firm drilling target, identified through detailed evaluation and, normally, 3D seismic data interpretation.

**Prospect map.** A map showing the locations of the areas where hydrocarbon generation is expected to have taken place and where potential traps are present.

**Proved reserves.** Quantities of oil or gas in existing fields that can be produced from existing wells.

**PVT.** Pressure–volume–temperature analysis. Laboratory procedure to determine the *formation volume factor.*

**P-wave.** Compressional wave.

**R**

**Recovery factor.** Abbreviated to R, it refers to the recoverable fraction of hydrocarbons initially in place, normally expressed as a percentage.

**Regression.** A relative fall in sea level, resulting in a withdrawal of the sea from the land.

**Relative time.** The age of one geological feature or event with respect to another.

**Reserves.** Portion of an identified resource that can be produced economically through existing technology.

**Reservoir rock.** In the context of conventional hydrocarbons, a reservoir is a permeable rock in which the hydrocarbons are stored and from which they may be extracted. It is usually in communication with a mature source bed.

**Resource.** A broad term, encompassing accumulations of all industrially useful materials which include mineral deposits as well as petroleum.

**Reverse fault.** A fault along which the hanging wall has moved up with respect to the footwall. Caused by compression.

**Risk analysis.** Assessment of exploration risk at a prospect level.

**Rock salt.** See *evaporite*.

## S

**Salt dome.** A bulge caused by an upward movement and intrusion of evaporitic rocks into overlying layers. Synonymous with *salt diapir*.

**Salt pillow.** A bulge formed by the accumulation of evaporitic material; a non-piercement salt dome.

**Sandstone.** A clastic sedimentary rock, formed by the consolidation of individual sand size particles ($1/16-2$ mm in diameter).

**Seal.** Impervious beds which lie on top of or adjacent to reservoirs and prevent further upward or lateral hydrocarbon movement. Evaporites and shales are the most common seals.

**Secondary migration.** Movement of hydrocarbons within the reservoir rock, leading to accumulation.

**Sedimentary basin.** A depression in the Earth's crust that contains a large thickness of *sedimentary rocks*.

**Sedimentary environment.** The physical and chemical conditions that prevail in an area where *sedimentary rocks* are being deposited.

**Sedimentary rocks.** Formed either by the cementing together or consolidation of fragments (grains) derived from pre-existing rocks, by direct precipitation from water or from the life processes of animals and plants.

**Seismic stratigraphy.** Use of the strength, character and patterns of the main reflections to deduce the nature of the rocks and reconstruct their environment of deposition.

**Sequence stratigraphy.** The study of packages of sediment composed of a relatively conformable succession of strata bounded above and below by regional unconformities. The objective is to construct models of deposition in terms of rises and falls in sea level.

**Shale.** A laminated, very fine grained clastic sedimentary rock composed of silt and clay size particles (less than 1/16 mm in diameter).

**Shale gas.** Gas produced from shale by a combination of horizontal drilling and hydraulic fracturing (fracking). Also known as tight gas.

**Shale oil.** Oil produced from shale by a combination of horizontal drilling and hydraulic fracturing (fracking). Also referred to as *tight oil.*

**Sill.** An intrusive igneous rock that invades pre-existing rocks, forming horizontal bodies.

**Siltstone.** A fine grained sedimentary rock composed of silt size particles (1/16−1/256 mm in diameter).

**Solution gas drive.** Production mechanism resulting from gas bubbling out of oil as reservoir pressure declines, expanding, forcing the oil out of the pores towards and up the production wells and ultimately to the surface.

**Source rock.** A fine grained sediment that in its natural setting has generated and released sufficient quantities of hydrocarbons to form a commercially viable conventional accumulation of oil and gas. Source rocks are clay or carbonate organic rich muds deposited under low energy, anoxic conditions. To function as a source rock, the sediment must contain a minimum of 2% by weight of organic matter.

**SPE.** Society of Petroleum Engineers. Constituted in 1957, it is a professional association with more than 143,000 members worldwide. Its objective is to advance technical knowledge concerning the exploration, development and production of oil and gas resources through publications, conferences, workshops, forums, educational courses and by promoting research.

**Spill point.** The lowest level at which hydrocarbons can be retained in the trap. Once a trap has been filled to its spill point, its hydrocarbon storage capacity ends. Any more hydrocarbons moving into the trap will spill out and continue to migrate until they encounter another trap along their migration path.

**Stacking.** Adding together signals from different receivers to upgrade the quality of seismic data.

**Step-out well.** See *delineation well.*

**STOIIP.** Stock tank oil initially in place.

**Stratigraphic trap.** Formed by variations in the lateral continuity of the reservoir, i.e. a change in facies.

**Stratigraphy.** An important branch of geology concerned with the study of the age and description of stratified rocks.

**Strike.** The trend of a dipping surface such as a bedding or fault plane. It is perpendicular to dip.

**Strike-slip fault.** A fault along which the movement is predominantly horizontal. Also known as a *transcurrent fault.*

**Structural geology.** The study of the deformation of rocks. Manifestations of deformation range from tilting and arching to large- and small-scale faults and thrusts.

**Structural trap.** Formed by changes in the shape of the reservoir or by faulting, throwing the reservoir against an impermeable layer across the fault plane.

**Structure contour map.** A map showing the variation in the shape of a surface, often the contact between two different rock types, with respect to a datum, usually sea level.

**Subduction.** The process of one lithospheric plate underthrusting another. See also *destructive plate margin.*

**Subsidence.** Sinking of the basin floor over a long period of time (tens of millions of years), creating *accommodation space* which in turn leads to sedimentation.

**S-wave.** Shear wave.

**Syncline.** A fold in which the limbs dip towards one another. In the case of a syncline, the 'core' is younger than the 'envelope'.

**T**

**TCF.** Trillion cubic feet.

**TCM.** Trillion cubic metres.

**Tectonics.** The processes that cause large-scale deformation of the Earth's crust, manifested by the formation of folds, faults, thrusts and mountain belts.

**Thrust.** A low angle reverse fault along which there has been a great deal of movement.

**TOC.** Total organic carbon.

**Tight gas.** Gas contained in low permeability formations. Used by some authors synonymously with shale gas.

**Tight oil.** Light crude oil contained in low permeability formations. Used by some authors synonymously with shale oil.

**Tilted fault block.** A rotated horst.

**Transcurrent fault.** See *strike-slip fault*.

**Transform fault plate margin.** A margin where the plates slip by each other and crust is neither created nor destroyed.

**Transgression.** A relative rise in sea level, resulting in an advance of the sea over the land.

**Trap.** A special situation in the reservoir that arrests the migration process and causes the hydrocarbons to accumulate. Trap formation must predate hydrocarbon migration.

**TWT.** Two way time.

**U**

**Unconformity.** A surface separating older from younger rocks and indicating a gap or interruption in the sedimentary record. It represents a time interval during which there was no deposition or the rocks were subjected to erosion. In some cases there is an angular difference between the attitude of the beds above and those below the surface of separation, in which case it is referred to as an angular unconformity.

**Unconventional resources.** Oil and gas deposits that are not stored in pore spaces in permeable rocks (the reservoir) and are not commercially recoverable by *conventional* drilling and production methods.

**Undiscovered reserves.** Quantities of oil and gas that could be found in an area.

**Uniformitarian principle.** The present is the key to the past, i.e. the processes that operate at present also operated in the past and produced the same results.

**USGS.** United States Geological Survey. Founded in 1879, USGS is a scientific agency of the American government. It is a fact-finding research organisation with no regulatory responsibility and has four major disciplines: biology, geography, geology and hydrology. The organisation produces a wide variety of publications, which include a bulletin, detailed topographical, geological, geophysical, hydrological and fossil fuel resources (coal oil and gas) maps of the United States. Periodically, USGS publishes estimates of the undiscovered conventional global oil and gas reserves. It has a staff of 8,760 and the largest earth sciences library in the world.

**V**

$\mathbf{V}_p$. Compressional wave velocity.
$\mathbf{V}_s$. Shear wave velocity.

**W**

**Water drive.** Production mechanism whereby water from the aquifer below the oil column moves up and occupies the pores vacated by oil or gas as the result of extraction. The OWC is pushed upwards, forcing the oil through the reservoir into the production wells and ultimately to the surface.

**Wildcat well.** See *Exploration well.*

# Index